Advanced Biopolymer-Based Nanocomposites and Hybrid Materials

Advanced Biopolymer-Based Nanocomposites and Hybrid Materials

Editors

**Armando J. D. Silvestre
Carmen S.R. Freire
Carla Vilela**

MDPI • Basel • Beijing • Wuhan • Barcelona • Belgrade • Manchester • Tokyo • Cluj • Tianjin

Editors
Armando J. D. Silvestre
CICECO—Aveiro Institute of Materials, Department of Chemistry, University of Aveiro Portugal

Carmen S.R. Freire
CICECO—Aveiro Institute of Materials, Department of Chemistry, University of Aveiro Portugal

Carla Vilela
CICECO—Aveiro Institute of Materials, Department of Chemistry, University of Aveiro Portugal

Editorial Office
MDPI
St. Alban-Anlage 66
4052 Basel, Switzerland

This is a reprint of articles from the Special Issue published online in the open access journal *Materials* (ISSN 1996-1944) (available at: https://www.mdpi.com/journal/materials/special_issues/biopolymer_nanocomposites_hybrid_materials).

For citation purposes, cite each article independently as indicated on the article page online and as indicated below:

LastName, A.A.; LastName, B.B.; LastName, C.C. Article Title. *Journal Name* **Year**, *Volume Number*, Page Range.

ISBN 978-3-0365-0510-7 (Hbk)
ISBN 978-3-0365-0511-4 (PDF)

© 2021 by the authors. Articles in this book are Open Access and distributed under the Creative Commons Attribution (CC BY) license, which allows users to download, copy and build upon published articles, as long as the author and publisher are properly credited, which ensures maximum dissemination and a wider impact of our publications.

The book as a whole is distributed by MDPI under the terms and conditions of the Creative Commons license CC BY-NC-ND.

Contents

About the Editors vii

Armando J. D. Silvestre, Carmen S. R. Freire and Carla Vilela
Special Issue: Advanced Biopolymer-Based Nanocomposites and Hybrid Materials
Reprinted from: *Materials* **2021**, *14*, 493, doi:10.3390/ma14030493 1

Carla Vilela, Catarina Moreirinha, Adelaide Almeida, Armando J. D. Silvestre and Carmen S. R. Freire
Zwitterionic Nanocellulose-Based Membranes for Organic Dye Removal
Reprinted from: *Materials* **2019**, *12*, 1404, doi:10.3390/ma12091404 5

Ricardo J. B. Pinto, Daniela Bispo, Carla Vilela, Alexandre M. P. Botas, Rute A. S. Ferreira, Ana C. Menezes, Fábio Campos, Helena Oliveira, Maria H. Abreu, Sónia A. O. Santos and Carmen S. R. Freire
One-Minute Synthesis of Size-Controlled Fucoidan- Gold Nanosystems: Antitumoral Activity and Dark Field Imaging
Reprinted from: *Materials* **2020**, *13*, 1076, doi:10.3390/ma13051076 21

Alexander N. Mitropoulos, F. John Burpo, Chi K. Nguyen, Enoch A. Nagelli, Madeline Y. Ryu, Jenny Wang, R. Kenneth Sims, Kamil Woronowicz and J. Kenneth Wickiser
Noble Metal Composite Porous Silk Fibroin Aerogel Fibers
Reprinted from: *Materials* **2019**, *12*, 894, doi:10.3390/ma12060894 35

Tze-Wen Chung, Weng-Pin Chen, Pei-Wen Tai, Hsin-Yu Lo and Ting-Ya Wu
Roles of Silk Fibroin on Characteristics of Hyaluronic Acid/Silk Fibroin Hydrogels for Tissue Engineering of Nucleus Pulposus
Reprinted from: *Materials* **2020**, *13*, 2750, doi:10.3390/ma13122750 47

Evaldas Balčiūnas, Nadežda Dreižė, Monika Grubliauskaitė, Silvija Urnikytė, Egidijus Šimoliūnas, Virginija Bukelskienė, Mindaugas Valius, Sara J. Baldock, John G. Hardy and Daiva Baltriukienė
Biocompatibility Investigation of Hybrid Organometallic Polymers for Sub-Micron 3D Printing via Laser Two-Photon Polymerisation
Reprinted from: *Materials* **2019**, *12*, 3932, doi:10.3390/ma12233932 65

Mihaela Tanase-Opedal, Eduardo Espinosa, Alejandro Rodríguez and Gary Chinga-Carrasco
Lignin: A Biopolymer from Forestry Biomass for Biocomposites and 3D Printing
Reprinted from: *Materials* **2019**, *12*, 3006, doi:10.3390/ma12183006 81

About the Editors

Armando J. D. Silvestre is a Full Professor and the Head of the Department of Chemistry of the University of Aveiro (Portugal). His research interests are focused on the biorefinery concept development, namely through the sustainable extraction and upgrading of added-value compounds from biomass addressing mainly bioactive compounds through the use of benign solvents (using ionic liquids, deep eutectic solvents, and supercritical CO_2); new biobased polymers derived from 2,5-furandicarboxylic acid and new functional (nano)composite materials and biomaterials based on biopolymers and cellulose (nano)fibers. He has published 1 book, 24 book chapters, 4 patents, 350+ papers, h-index 57, 11,000+ citations.

Carmen S.R. Freire is a Principal Researcher at CICECO—Aveiro Institute of Materials (University of Aveiro, Portugal) in the Biorefinery and bio-based materials research area. Her research interests include the production and application of biogenic nanofibers (bacterial cellulose and protein fibrils), nanostructured biocomposites and hybrid materials; bio-based materials for biomedical applications (wound healing, drug delivery, and 3D bioprinting), biocomposites and functional paper materials; chemical modification of (nano)cellulose fibers and other polysaccharides and their characterization and applications. C. Freire is the author of more than 220 papers (with 7500+ citations, h-index 52), 5 patents, 9 book chapters, and more than 180 communications in international and national conferences.

Carla Vilela is an assistant researcher at CICECO—Aveiro Institute of Materials (University of Aveiro, Portugal) and her main research interest includes the sustainable use of biopolymers, namely polysaccharides (e.g., nanocellulose, pullulan, chitosan) and proteins (e.g., gelatin, lysozyme), for the development of novel functional nanostructured materials for both technological (e.g., active food packaging and fuel cells) and biomedical (e.g., drug delivery and wound healing) applications. She is the author of 62 SCI papers (h-index: 22; citations: 1600+), 1 book, 5 book chapters, 5 conference proceedings, and more than 90 communications in national and international conferences.

Editorial

Special Issue: Advanced Biopolymer-Based Nanocomposites and Hybrid Materials

Armando J. D. Silvestre, Carmen S. R. Freire and Carla Vilela *

CICECO—Department of Chemistry, Aveiro Institute of Materials, University of Aveiro, 3810-193 Aveiro, Portugal; armsil@ua.pt (A.J.D.S.); cfreire@ua.pt (C.S.R.F.)
* Correspondence: cvilela@ua.pt

Received: 11 January 2021; Accepted: 18 January 2021; Published: 21 January 2021

The gamut of natural polymers, from polysaccharides to proteins, exhibit peculiar features and multiple functionalities that are being exploited to engineer advanced nanocomposites and hybrid materials. In fact, the growing environmental concerns explain the focus of the increasing scientific activity on both polysaccharides (e.g., cellulose, chitosan, alginate, fucoidan, and hyaluronic acid) and proteins (e.g., silk, collagen, and gelatin), given their remarkable potential for the design of all categories of materials for application in multiple fields, for example, in electronics, energy, environment, biology, and medicine [1,2]. This Special Issue of *Materials*—belonging to the section *Biomaterials*—contains a collection of six research papers about advanced biopolymer-based nanocomposites and hybrid materials. The collected papers made use of biopolymers, such as nanocellulose [3], fucoidan [4], silk fibroin [5,6], hyaluronic acid [6], and lignin [7], in combination with a polyzwitterion [3], metal nanoparticles [4,5], organometallic polymers [8], and a thermoplastic polymer [7], to develop advanced systems for water remediation [3], cancer treatment [4], catalysis, sensing, and energy storage [5], tissue engineering [6,8], and 3D printing [7,8].

The investigation of Vilela et al. [3] demonstrated that zwitterionic nanocomposite membranes comprising bacterial nanocellulose and cross-linked poly(2-methacryloyloxyethyl phosphorylcholine) can be used as tools for water remediation. The combination of the bacterial polysaccharide and the polyzwitterion originated robust nanocomposite adsorbent membranes with high water uptake capacity in different pH media and antimicrobial activity against Gram-positive (*Staphylococcus aureus*) and Gram-negative (*Escherichia coli*) pathogenic bacteria. Furthermore, the nanocellulose-based adsorbent membranes were capable of adsorbing two model ionic organic azo dyes, namely methylene blue and methyl orange, with toxicity towards humans and the environment, thus proving their potential in the context of water remediation.

In a different study, Pinto et al. [4] explored the potential of fucoidan-gold nanosystems for cancer therapy. The authors effectively developed gold nanoparticles (AuNPs) coated with fucoidan via one-minute microwave-assisted synthesis using a fucoidan-enriched fraction extracted from *Fucus vesiculosus* (species of brown algae), which played the simultaneous role of reducing and capping agents. This one-minute synthesis originated monodispersed and spherical fucoidan-AuNPs with antitumoral activity against three human tumor cell lines, namely MNT-1 (pigmented human melanoma cells), HepG2 (human hepatocyte carcinoma), and MG-63 (human osteosarcoma) cell lines. Moreover, the combination between flow cytometry and dark-field imaging confirmed the cellular uptake of the fucoidan-AuNPs by the MG-63 cell line.

As pointed out by Mitropoulos et al. [5], noble metal nanoparticles, namely palladium, platinum, and gold, can be anchored at the surface of silk fibroin (SF) to prepare composite aerogel fibers. The porous silk gels were produced by dissolving the SF in hexafluoro-2-propanol (HFIP), followed by casting in silicon tubes and physical crosslinking with ethanol. These gels were then equilibrated in the noble metal salt solutions, reduced with sodium borohydride, and dried via supercritical technology to produce the porous aerogel fibers coated with noble metal nanoparticles, for application in catalysis, energy storage, and conversion.

Chung et al. [6] delve into the effect of SF on the characteristics of hyaluronic acid (HA)/silk fibroin hydrogels for tissue engineering of nucleus pulposus (NP). The authors concluded that the strain of SF and the weight ratios of SF to HA influenced the rheological properties of the crosslinked silk fibroin/hyaluronic acid hydrogels. Additionally, the biocompatible SF/HA hydrogels favored the differentiation of human bone marrow-derived mesenchymal stem cell (hBMSC) to NP cells, which demonstrated the suitability of these hydrogels in tissue engineering for NP regeneration.

In another study, Balčiūnas et al. [8] investigated the biocompatibility of hybrid organometallic polymers for sub-micron 3D printing via laser two-photon polymerization. The Al and Zr containing hybrid organometallic polymers supported cellular growth to full confluency and promoted collagen synthesis. Therefore, these hybrid organometallic polymers can be an asset in constructing the tissue-engineered grafts of the future.

Finally, Tanase-Opedal et al. [7] manufactured composite filaments by compounding the thermoplastic poly(lactic acid) (PLA) with lignin obtained from Spruce biomass via the soda pulping process. The aromatic natural polymer from forestry biomass acted as a nucleating agent, which promoted further crystallization of PLA, and originated composite materials with antioxidant potential for application in 3D printing.

Overall, the six papers of this Special Issue of *Materials* covered some examples of materials resulting from the combination of natural polymers with distinct partners (e.g., synthetic polymers or metal nanoparticles), and contributed to the development of advanced nanocomposites and hybrid materials for both technological and biomedical applications.

Funding: This work was developed within the scope of the project CICECO-Aveiro Institute of Materials (UIDB/50011/2020 & UIDP/50011/2020), financed by national funds through the Portuguese Foundation for Science and Technology (FCT)/MCTES. FCT is also acknowledged for the research contract under Scientific Employment Stimulus to C.V. (CEECIND/00263/2018).

Acknowledgments: The Guest Editors wish to acknowledge the authors for their vital contributions to this Special Issue, the reviewers for their hard work in reviewing the submitted papers, and the editorial staff of *Materials* for their extraordinary support.

Conflicts of Interest: The authors declare no conflict of interest.

References

1. Vilela, C.; Pinto, R.J.B.; Pinto, S.; Marques, P.A.A.P.; Silvestre, A.J.D.; Freire, C.S.R. *Polysaccharide Based Hybrid Materials: Metals And Metal Oxides, Graphene And Carbon Nanotubes*, 1st ed.; Springer: Berlin, Germany, 2018; ISBN 978-3-030-00346-3.
2. Silva, N.H.C.S.; Vilela, C.; Marrucho, I.M.; Freire, C.S.R.; Pascoal Neto, C.; Silvestre, A.J.D. Protein-based materials: From sources to innovative sustainable materials for biomedical applications. *J. Mater. Chem. B* **2014**, *2*, 3715–3740. [CrossRef] [PubMed]
3. Vilela, C.; Moreirinha, C.; Almeida, A.; Silvestre, A.J.D.; Freire, C.S.R. Zwitterionic nanocellulose-based membranes for organic dye removal. *Materials* **2019**, *12*, 1404. [CrossRef] [PubMed]
4. Pinto, R.J.B.; Bispo, D.; Vilela, C.; Botas, A.M.P.; Ferrreira, R.A.S.; Menezes, A.C.; Campos, F.; Oliveira, H.; Abreu, M.H.; Santos, S.A.O.; et al. One-minute synthesis of size-controlled fucoidan-gold nanosystems: Antitumoral activity and dark field imaging. *Materials* **2020**, *13*, 1076. [CrossRef] [PubMed]
5. Mitropoulos, A.N.; Burpo, F.J.; Nguyen, C.K.; Nagelli, E.A.; Ryu, M.Y.; Wang, J.; Sims, R.K.; Woronowicz, K.; Wickiser, J.K. Noble metal composite porous silk fibroin aerogel fibers. *Materials* **2019**, *12*, 894. [CrossRef] [PubMed]
6. Chung, T.W.; Chen, W.P.; Tai, P.W.; Lo, H.Y.; Wu, T.Y. Roles of silk fibroin on characteristics of hyaluronic acid/silk fibroin hydrogels for tissue engineering of nucleus pulposus. *Materials* **2020**, *13*, 2750. [CrossRef] [PubMed]
7. Tanase-Opedal, M.; Espinosa, E.; Rodríguez, A.; Chinga-Carrasco, G. Lignin: A biopolymer from forestry biomass for biocomposites and 3D printing. *Materials* **2019**, *12*, 3006. [CrossRef] [PubMed]

8. Balčiunas, E.; Dreiže, N.; Grubliauskaite, M.; Urnikyte, S.; Šimoliunas, E.; Bukelskiene, V.; Valius, M.; Baldock, S.J.; Hardy, J.G.; Baltriukiene, D. Biocompatibility investigation of hybrid organometallic polymers for sub-micron 3D printing via laser two-photon polymerisation. *Materials* **2019**, *12*, 3932. [CrossRef] [PubMed]

Publisher's Note: MDPI stays neutral with regard to jurisdictional claims in published maps and institutional affiliations.

© 2021 by the authors. Licensee MDPI, Basel, Switzerland. This article is an open access article distributed under the terms and conditions of the Creative Commons Attribution (CC BY) license (http://creativecommons.org/licenses/by/4.0/).

Article

Zwitterionic Nanocellulose-Based Membranes for Organic Dye Removal

Carla Vilela [1,*], Catarina Moreirinha [1], Adelaide Almeida [2], Armando J. D. Silvestre [1] and Carmen S. R. Freire [1]

[1] Department of Chemistry, CICECO – Aveiro Institute of Materials, University of Aveiro, 3810-193 Aveiro, Portugal; catarina.fm@ua.pt (C.M.); armsil@ua.pt (A.J.D.S.); cfreire@ua.pt (C.S.R.F.)
[2] Department of Biology and CESAM, University of Aveiro, 3810-193 Aveiro, Portugal; aalmeida@ua.pt
* Correspondence: cvilela@ua.pt

Received: 10 April 2019; Accepted: 24 April 2019; Published: 30 April 2019

Abstract: The development of efficient and environmentally-friendly nanomaterials to remove contaminants and pollutants (including harmful organic dyes) ravaging water sources is of major importance. Herein, zwitterionic nanocomposite membranes consisting of cross-linked poly(2-methacryloyloxyethyl phosphorylcholine) (PMPC) and bacterial nanocellulose (BNC) were prepared and tested as tools for water remediation. These nanocomposite membranes fabricated via the one-pot polymerization of the zwitterionic monomer, 2-methacryloyloxyethyl phosphorylcholine, within the BNC three-dimensional porous network, exhibit thermal stability up to 250 °C, good mechanical performance (Young's modulus ≥ 430 MPa) and high water-uptake capacity (627%–912%) in different pH media. Moreover, these zwitterionic membranes reduced the bacterial concentration of both gram-positive (*Staphylococcus aureus*) and gram-negative (*Escherichia coli*) pathogenic bacteria with maxima of 4.3– and 1.8–log CFU reduction, respectively, which might be a major advantage in reducing or avoiding bacterial growth in contaminated water. The removal of two water-soluble model dyes, namely methylene blue (MB, cationic) and methyl orange (MO, anionic), from water was also assessed and the results demonstrated that both dyes were successfully removed under the studied conditions, reaching a maximum of ionic dye adsorption of *ca*. 4.4–4.5 mg g^{-1}. This combination of properties provides these PMPC/BNC nanocomposites with potential for application as antibacterial bio-based adsorbent membranes for water remediation of anionic and cationic dyes.

Keywords: bacterial nanocellulose; poly(2-methacryloyloxyethyl phosphorylcholine); zwitterionic nanocomposites; dye removal; water remediation; antibacterial activity

---

1. Introduction

The need for water remediation systems designed to eliminate contaminants and pollutants ravaging water sources is a global problem and, thus, is part of the goals of the 2030 Agenda for Sustainable Development [1]. Nevertheless, the struggle to remove heavy metal ions, pesticides and other dissolved organic pollutants is a difficult war and some of the efforts of researchers to accomplish such a target are directed towards the development of environmentally friendly porous nanomaterials [2,3]. In fact, big bets are being placed on systems derived from naturally occurring polymers, such as polysaccharides [4,5]. Within the portfolio of commended biopolymers, cellulose and its nanoscale forms, namely cellulose nanocrystals (CNCs), cellulose nanofibrils (CNFs) and bacterial nanocellulose (BNC), show tremendous potential for environmental and water remediation as recently reviewed [6–8]. The high interest in the latter nanocellulose, *viz*. the exopolysaccharide BNC biosynthesized by some non-pathogenic bacterial strains [9], lies in its 3D structure with an ultrafine network of physically entangled cellulose nanofibers, which is responsible for its *in-situ* moldability, shape retention, inherent biodegradability, high water-holding capacity and porous structure [10,11].

BNC, its derivatives and composites [12] have already been used in the manufacture of water remediation systems for the removal of dyes [13,14], oil [15] and heavy metals [16,17]. To the best of our knowledge, the partnership between BNC and a zwitterionic polymer has never been explored for the fabrication of nanocomposites, aiming at the simultaneous removal of anionic and cationic organic dyes. Under these premises, 2-methacryloyloxyethyl phosphorylcholine (MPC) was selected as a non-toxic polymerizable monomer due to its methacrylic functional group and zwitterionic phosphorylcholine moiety, consisting of a phosphate anion and a trimethylammonium cation [18], which are prone to establish electrostatic interactions with positively and negatively charged molecules. Furthermore, the unique hydration state [18], antimicrobial, bioinert and antifouling properties of MPC polymer [19–21] will be a major asset in reducing or avoiding bacterial growth in contaminated water.

The present study contemplates the fabrication of nanocomposite membranes consisting of cross-linked poly(2-methacryloyloxyethyl phosphorylcholine) (PMPC) and BNC via the one-pot polymerization of the corresponding non-toxic zwitterionic monomer (*i.e.* MPC) within the BNC three-dimensional porous network. The structure, morphology, thermal stability, mechanical properties, antibacterial activity towards *Staphylococcus aureus* and *Escherichia coli*, water-uptake capacity and removal of cationic and anionic organic dyes of the resulting nanocomposites were assessed.

2. Materials and Methods

2.1. Chemicals, materials and microorganisms

2-Methacryloyloxyethyl phosphorylcholine (MPC, 97%), 2,2'-azobis(2-methylpropionamidine) dihydrochloride (AAPH, 97%), *N,N'*-methylenebis(acrylamide) (MBA, 99%), methylene blue (MB, dye content ≥ 82%), methyl orange (MO, dye content 85 %) and paraffin oil (puriss., 0.827–0.890 g mL^{-1} at 20 °C) were purchased from Sigma-Aldrich (Sintra, Portugal) and used as received. Ultrapure water (Type 1, 18.2 MX·cm at 25 °C) was obtained from a Simplicity® Water Purification System (Merck, Darmstadt, Germany). Other chemicals and solvents were of laboratory grade.

Bacterial nanocellulose (BNC) wet membranes were biosynthesized in our laboratory using the *Gluconacetobacter sacchari* bacterial strain [22]. *Staphylococcus aureus* (ATCC 6538) and *Escherichia coli* (ATCC 25922) was provided by DSMZ – Deutsche Sammlung von Mikroorganismen und Zellkulturen GmbH (German Collection of Microorganisms and Cell Cultures).

2.2. Preparation of PMPC/BNC nanocomposites

Wet BNC membranes (diameter: *ca.* 70 mm) with 40% water content were placed in stoppered glass-reactors and purged with nitrogen. Simultaneously, aqueous solutions of monomer (MPC, 1:3 and 1:5 BNC/MPC weight fraction), cross-linker (MBA, 5.0% w/w relative to monomer), and radical initiator (AAPH, 1.0% w/w relative to monomer) were prepared and transferred to the glass-reactors containing the drained BNC membranes. After the complete incorporation of the corresponding solution into the BNC membrane, during 1 h in ice, the reaction mixtures were heated in an oil bath at 70 °C and left to react for 6 h. Afterwards, the nanocomposites were repeatedly washed with water, oven dried at 40 °C for 12 h, and kept in a desiccator until further use. All experiments were made in triplicate and samples of cross-linked PMPC were also prepared in the absence of BNC for comparison.

2.3. Characterization methods

2.3.1. Thickness

A hand-held digital micrometer (Mitutoyo Corporation, Tokyo, Japan) with an accuracy of 0.001 mm was used to measure the thickness of the membranes. All measurements were randomly performed at different sites of the membranes.

2.3.2. Ultraviolet-visible spectroscopy (UV–vis)

The transmittance spectra of the samples were acquired with a Shimadzu UV-1800 UV-Vis spectrophotometer (Shimadzu Corp., Kyoto, Japan) equipped with a quartz window plate with 10 mm diameter, bearing the holder in the vertical position. Spectra were recorded at room temperature (RT) in steps of 1 nm in the range 250–700 nm.

2.3.3. Attenuated total reflection-Fourier transform Infrared (ATR-FTIR)

ATR-FTIR spectra were recorded with a Perkin-Elmer FT-IR System Spectrum BX spectrophotometer (Perkin-Elmer Inc., Massachusetts, USA) equipped with a single horizontal Golden Gate ATR cell, over the range of 600–4000 cm^{-1} at a resolution of 4 cm^{-1} over 32 scans.

2.3.4. Solid-state carbon cross-polarization/magic-angle-spinning nuclear magnetic resonance (^{13}C CP/MAS NMR)

^{13}C CP/MAS NMR spectra were collected on a Bruker Avance III 400 spectrometer (Bruker Corporation, Massachusetts, USA) operating at a B0 field of 9.4 T using 9 kHz MAS with proton 90° pulse of 3 µs, time between scans of 3 s, and a contact time of 2000 µs. ^{13}C chemical shifts were referenced to glycine (C=O at δ 176 ppm).

2.3.5. X-ray diffraction (XRD)

XRD was performed on a Phillips X'pert MPD diffractometer (PANalytical, Eindhoven, Netherlands) using Cu Kα radiation (λ = 1.541 Å) with a scan rate of 0.05° s^{-1}. The XRD patterns were collected in reflection mode with the membranes placed on a Si wafer (negligible background signal) for mechanical support and thus avoid sample bending.

2.3.6. Scanning electron microscopy (SEM) coupled with energy dispersive X-ray spectroscopy (EDS)

SEM images of the cross-section of the membranes were obtained with a HR-SEM-SE SU-70 Hitachi microscope (Hitachi High-Technologies Corporation, Tokyo, Japan) operating at 4 kV. The microscope was equipped with an EDS Bruker QUANTAX 400 detector for elemental analysis. The samples were fractured in liquid nitrogen, placed on a steel plate and coated with a carbon film prior to analysis.

2.3.7. Thermogravimetric analysis (TGA)

TGA was carried out with a SETSYS Setaram TGA analyzer (SETARAM Instrumentation, Lyon France) equipped with a platinum cell. The samples were heated from RT to 800 °C at a constant rate of 10 °C min^{-1} under a nitrogen atmosphere (200 mL min^{-1}).

2.3.8. Tensile tests

Tensile tests were performed on a uniaxial Instron 5564 testing machine (Instron Corporation, Maryland, USA) in the traction mode at a cross-head velocity of 10 mm min^{-1} using a 500 N static load cell. The specimens were rectangular strips (50 × 10 mm^2) previously dried at 40 °C and equilibrated at RT in a 50% relative humidity (RH) atmosphere prior to testing. All measurements were performed on five replicates and the results were expressed as the average value.

2.3.9. Water-uptake capacity

The water-uptake of the nanocomposites under different pH conditions was determined via immersion of dry specimens with 10 × 10 mm^2 in aqueous solutions of 0.01 M HCl (pH 2.1), phosphate buffer saline (pH 7.4) and 0.01 M NaOH (pH 12) at RT for 48 h. After removing the specimens out of the respective medium, the wet surfaces were dried in filter paper, and the wet weight (W_w) was measured. The water-uptake is calculated by the equation: $W_{uptake}(\%) = (W_w - W_0) \times W_0^{-1} \times 100$, where W_0 is the initial weight of the dry membrane.

2.4. In vitro antibacterial activity

The antibacterial activity of the nanocomposite membranes was tested against S. aureus and E. coli. The bacterial pre-inoculum cultures were grown for 24 h in tryptic soy broth (TSB) growth medium at 37 °C under shaking at 120 rpm. Before the assay, the density of the bacterial culture was adjusted to 0.5 McFarland in phosphate buffered saline (PBS) solution (pH 7.4) to obtain a bacterial concentration of 10^8 to 10^9 colony forming units per mL (CFU mL^{-1}). Each membrane sample (50 × 50 mm^2) was placed in contact with 5 mL of bacterial suspension. A bacteria cell suspension was tested as the control and BNC was tested as a blank reference. All samples were incubated at 37 °C under horizontal shaking at 120 rpm. At 24 h contact time, aliquots (100 µL) of each sample and controls were collected and the bacteria cell concentration (CFU mL^{-1}) was determined by plating serial dilutions on tryptic soy agar (TSA) medium. The plates were incubated at 37 °C for 24 h. The CFU were determined on the most appropriate dilution on the agar plates. Three independent experiments were carried out and, for each, two replicates were plated. The bacteria reduction of the samples was calculated as follows: $log\ reduction = log\ CFU_{control} - log\ CFU_{membrane}$.

2.5. Dye removal capacity

The dye removal capacity of the PMPC/BNC nanocomposite membranes was evaluated by immersing dry samples (20 × 20 mm^2) in 25 mL of methyl blue (MB) and methyl orange (MO) aqueous solutions (25 mg L^{-1}, pH 5.7), and stirred (200 rpm) for 12 h at RT. Then, the membranes were removed from the solution and the residual concentration of dye determined by UV–vis spectroscopy (Shimadzu UV-1800 UV-Vis spectrophotometer, Kyoto, Japan) at 655 nm for MB and 463 nm for MO. Linear calibration curves for each dye in the range 0.4–3.1 µg mL^{-1} were obtained: $y = 0.1919x + 0.0014$ ($R^2 = 0.9991$) for MB and $y = 0.0881x - 0.0046$ ($R^2 = 0.9992$) for MO. The dye removal amount (mg g^{-1}) was calculated by: $q = (C_i - C_t) \times W^{-1} \times V$, where C_i is the initial dye concentration (mg L^{-1}), C_t is the dye concentration at time t (h), W is the weight (g) of the membrane and V is the volume (L) of the dye solution.

Additionally, a dry sample of PMPC/BNC_2 nanocomposite (20 × 20 mm^2) was immersed in 25 mL of paraffin oil containing 1 mL of MB and MO aqueous solutions (25 mg L^{-1}).

2.6. Statistical analysis

Statistical significance was determined using an analysis of variance (ANOVA) and Tukey's test (OriginPro, version 9.0.0, OriginLab Corporation, Northampton, MA, USA). Statistical significance was established at $p < 0.05$.

3. Results and Discussion

3.1. PMPC/BNC nanocomposites: preparation and characterization

The one-pot in-situ free radical polymerization of MPC, inside the swollen BNC network and using MBA as cross-linker, was used to produce two PMPC/BNC nanocomposites (Figure 1) with distinct compositions (Table 1). The cross-linker was chosen based on former studies [23,24] and utilized with the goal of preserving the water-soluble zwitterionic homopolymer inside the BNC porous network during washing and utilization. The resulting nanocomposites contain 21 ± 3 wt.% and 46 ± 13 wt.% of BNC (W_{BNC}/W_{total}), and concomitantly 79 ± 3 wt.% and 54 ± 13 wt.% of PMPC (W_{PMPC}/W_{total}), which correspond to nanocomposite materials containing 479 ± 118 mg and 859 ± 90 mg of PMPC per cm^3 of membrane, respectively, as listed in Table 1. The thickness of the membranes increased from 92 ± 21 µm for neat BNC to 133 ± 65 µm for PMPC/BNC_1 and 226 ± 35 µm for PMPC/BNC_2 (Table 1) due to the inclusion of the cross-linked PMPC into the three-dimensional structure of BNC. The membranes are macroscopically homogeneous with no discernible irregularities on either side of the materials surfaces, indicating a good dispersion of the cross-linked PMPC polymer inside the BNC network. After the incorporation of PMPC into the BNC network, the transparency of the nanocomposites significantly increased, as displayed in Figure 1B and confirmed by transmittance values in the visible range (400–700 nm) of 58.1–65.6% for PMPC/BNC_1 and 60.5–68.2% for PMPC/BNC_2 (Figure 1C). In the ultraviolet region (200–400 nm), the transmittance remained below 5% until 265 nm for PMPC/BNC_1 and 250 nm for PMPC/BNC_2, and then steadily increased to 58.0% and 60.5% at 400 nm for PMPC/BNC_1 and PMPC/BNC_2, respectively. Furthermore, PMPC/BNC_2 presents higher values of transmittance and concomitantly lower absorbance values, which points to a transmittance augment with higher content of cross-linked PMPC (Figure 1C). An analogous trend was observed for other BNC-based nanocomposites containing for example polycaprolactone [25], poly(methacroylcholine chloride) [23] and polyaniline [26].

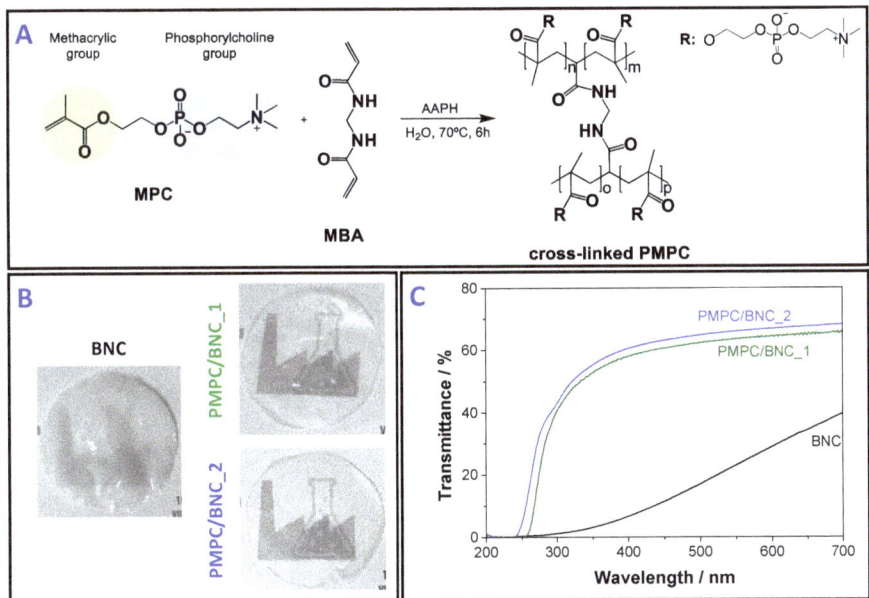

Figure 1. (**A**) Radical polymerization of MPC in the presence of MBA as cross-linker, yielding cross-linked PMPC, (**B**) photographs of neat BNC and nanocomposites PMPC/BNC_1 and PMPC/BNC_2, and (**C**) the corresponding UV-visible transmission spectra.

Table 1. List of the studied membranes with the respective weight compositions and thicknesses.

Samples	Composition [a]			Thickness/μm
	W_{BNC}/W_{total}	W_{PMPC}/W_{total}	W_{PMPC}/V_{total} (mg cm^{-3}) [b]	
BNC	1.0	–	–	92 ± 21
PMPC/BNC_1	0.46 ± 0.13	0.54 ± 0.13	479 ± 118	133 ± 65
PMPC/BNC_2	0.21 ± 0.03	0.79 ± 0.03	859 ± 90	226 ± 35

[a] The composition was calculated by considering the weight of the nanocomposite membrane (W_{total}), BNC (W_{BNC}) and PMPC cross-linked polymer ($W_{PMPC} = W_{total} - W_{BNC}$); [b] Ratio between the mass of the cross-linked PMPC (W_{PMPC}) and the volume of the nanocomposite membrane (V_{total}); all values are the mean of at least three replicates with the respective standard deviations.

The infrared spectra of neat BNC, cross-linked PMPC, and nanocomposites PMPC/BNC_1 and PMPC/BNC_2 are shown in Figure 2. The ATR-FTIR spectra of the PMPC/BNC membranes present the absorption bands of cellulose at 3340 cm^{-1} (O–H stretching), 2900 cm^{-1} (C–H stretching), 1310 cm^{-1} (O–H in plane bending) and 1030 cm^{-1} (C–O stretching) [27], jointly with those of the cross-linked PMPC at 1715 cm^{-1} (C=O stretching), 1479 cm^{-1} (N$^+$(CH$_3$)$_3$ bending), 1228 cm^{-1} (P=O stretching),

1056 cm^{-1} (P–O–C stretching) and 953 cm^{-1} (N$^+$(CH$_3$)$_3$ stretching) [28,29]. The presence of these absorption bands and the absence of one at about 1640 cm^{-1} corresponding to the C=C stretching of the methacrylic group of the starting monomer corroborated the occurrence of the *in-situ* free radical polymerization of MPC inside the BNC network. Furthermore, the relative intensity of the bands assigned to the cross-linked PMPC is in accordance with the W_{PMPC}/W_{total} ratio measured for each nanocomposite (Table 1).

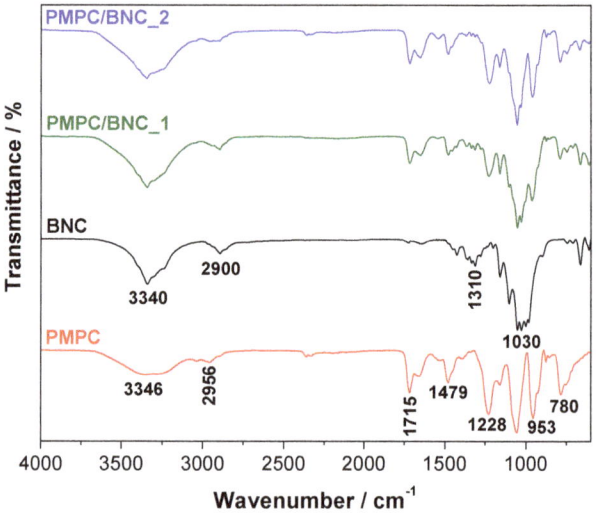

Figure 2. ATR-FTIR spectra of cross-linked PMPC, neat BNC, and nanocomposites PMPC/BNC_1 and PMPC/BNC_2.

The solid-state ^{13}C CP/MAS NMR spectra (Figure 3) of the membranes show the typical carbon resonances of cellulose at δ 65.2 ppm (C6), 71.6–74.5 ppm (C2,3,5), 88.9 ppm (C4) and 105.1 ppm (C1) [27], in combination with those of cross-linked PMPC at δ 18.4 ppm (CH$_3$ of polymer backbone), 44.9 ppm (quaternary C of polymer backbone), 54.3 ppm (CH$_2$ of polymer backbone and N$^+$(CH$_3$)$_3$), 59.7 ppm (OCH$_2$CH$_2$N$^+$(CH$_3$)$_3$), 65.5 ppm (OCH$_2$CH$_2$O and CH$_2$CH$_2$N$^+$(CH$_3$)$_3$) and 176.6 ppm (C=O). In addition, the truancy of the resonances allocated to the C=C double bond of the methacrylic group of the monomer [30] and cross-linker, supports their complete consumption during the polymerization and/or removal during the washing steps, as previously established by ATR-FTIR analysis.

The XRD patterns of the nanocomposites were compared with those of the individual components, namely cross-linked PMPC and BNC, to obtain an indication of the nanomaterials' crystallinity. Figure 4 shows the amorphous character of the cross-linked PMPC with a very broad band centered at 2θ ≈ 18°, and the crystalline nature of BNC with a diffraction pattern characteristic of cellulose I (native cellulose) composed of highly-ordered and least-ordered regions. The nanocomposites display a diffractogram with three peaks corresponding to the (101) plane at 2θ ≈ 14.7°, (10$\bar{1}$) plane at 2θ ≈ 16.8° and (002) plane at 2θ ≈ 22.8° [27], which are representative of the cellulosic substrate. The addition of the cross-linked PMPC is evident through the reduction of the peaks of the (101) and (10$\bar{1}$) planes, most likely linked to the augment of disordered cellulose domains due to the presence of the amorphous polymer. A comparable trend was reported for other BNC-based nanocomposites containing for instance poly(bis[2-(methacryloyloxy)ethyl] phosphate) [31] and poly(4-styrene sulfonic acid) [32].

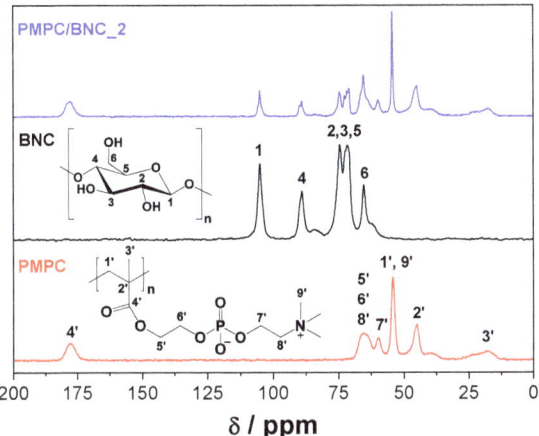

Figure 3. ^{13}C CP/MAS NMR spectra of cross-linked PMPC, neat BNC and nanocomposite PMPC/BNC_2.

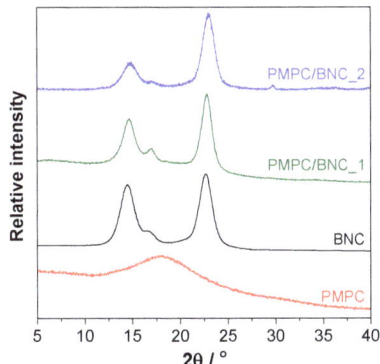

Figure 4. X-ray diffractograms of cross-linked PMPC, neat BNC, and nanocomposites PMPC/BNC_1 and PMPC/BNC_2.

SEM micrographs of the cross-section of neat BNC and nanocomposites PMPC/BNC_1 and PMPC/BNC_2 are compiled in Figure 5A. It is clearly visible that the lamellar microstructure of neat BNC disappeared in the nanocomposites due to the filling of the lamellar spaces by the cross-linked PMPC, particularly in the case of PMPC/BNC_2 with 859 ± 90 mg of PMPC *per* cm^3 of membrane. The SEM/EDS analysis reiterates the presence of PMPC and BNC through the detection of carbon (C), nitrogen (N), oxygen (O) and phosphorous (P) peaks at 0.27, 0.39, 0.51 and 2.01 keV, respectively, as illustrated in Figure 5B for PMPC/BNC_2. Moreover, the SEM/EDS mapping (Figure 5C) confirmed the uniform distribution of nitrogen and phosphorous of the cross-linked polymer within the BNC nanofibrous network, since both elements are only present in the zwitterionic PMPC.

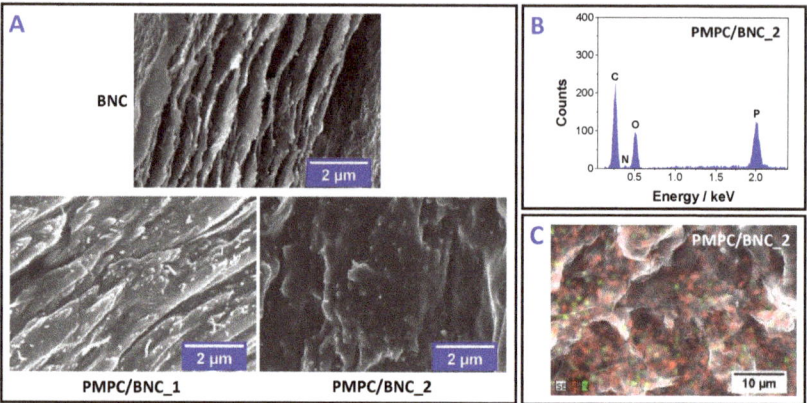

Figure 5. (**A**) SEM micrographs of the cross-section of neat BNC, and nanocomposites PMPC/BNC_1 and PMPC/BNC_2; EDS spectrum (**B**) and micrograph (**C**) for nitrogen and phosphorous elemental mapping of nanocomposite PMPC/BNC_2.

3.2. Thermal stability

TGA analysis was used to study the thermal stability of the PMPC/BNC nanocomposite membranes, as well as of their individual components. Figure 6A shows the thermograms of cross-linked PMPC and neat BNC, while Figure 6B presents the thermograms of PMPC/BNC_1 and PMPC/BNC_2 nanocomposites. The thermal degradation profile of the cross-linked PMPC is characterized by two consecutive steps (apart from the dehydration at about 100 °C with a loss of *ca.* 7.5%) with maximum decomposition temperatures of 284 °C (loss of *ca.* 21%) and 385 °C (loss of *ca.* 28%, Figure 6A) corresponding to the pyrolysis of the PMPC skeleton. The thermogram of BNC shows a single weight-loss step with initial and maximum decomposition temperatures of 290 °C and 342.5 °C (loss of *ca.* 68%, Figure 6A), respectively, allocated to the pyrolysis of the cellulose skeleton [33,34]. PMPC left a residue at 800 °C corresponding to about 34% of the initial mass, whereas BNC only left a residue of *ca.* 18% at the end of the analysis.

The thermal degradation profiles of both nanocomposites (Figure 6B) follow a double weight-loss step, aside from the water loss below 100 °C (loss of *ca.* 10%). PMPB/BNC_1 has maximum decomposition temperatures at 290 °C (loss of *ca.* 24%) and 382 °C (loss of *ca.* 17%) with a final residue of 35%, while for PMPC/BNC_2 the maximum occurs at 288 °C (loss of *ca.* 19%) and 383 °C (loss of *ca.* 20%) with a residue of 32% at the end of the analysis (800 °C). This two-step pathway is associated first with the simultaneous pyrolysis of cellulose and cross-linked PMPC, and the last stage with only the zwitterionic polymer. This pattern points to the reduction of the thermal stability of the nanocomposites when compared to neat BNC, as already described for other BNC-based nanocomposites containing polymers with lower thermal stability [35]. Even so, the two PMPC/BNC membranes exhibit good thermal stability up to 250 °C (Figure 6B).

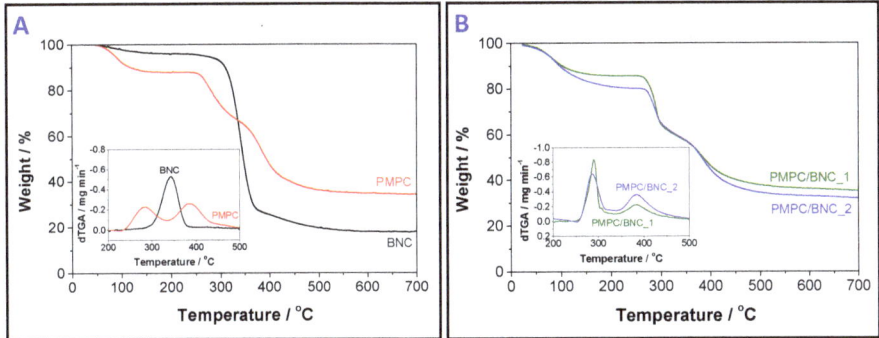

Figure 6. Thermograms of (**A**) cross-linked PMPC and neat BNC, and (**B**) membranes PMPC/BNC_1 and PMPC/BNC_2 under nitrogen atmosphere. The inset curves correspond to the derivative.

3.3. Mechanical properties

Figure 7 compiles the tensile tests data, namely in terms of Young's modulus, tensile strength and elongation at break, determined from the stress-strain curves. Although the tensile tests were not performed for the cross-linked PMPC due to its lack of film-forming aptitude, the cooperative effect between PMPC and BNC relies on the mechanical properties of both nanocomposites. Overall, the Young's modulus and tensile strength of the two membranes increased with the increasing content of the cellulosic substrate (Figure 7A,B), on account of its good mechanical performance, namely Young's modulus of 6.6 ± 1.8 GPa and tensile strength of 221 ± 48 MPa. In fact, the former parameter increased from 430 ± 150 MPa for PMPC/BNC_2 with 21 wt.% of BNC to 3.3 ± 0.8 GPa for PMPC/BNC_1 with 46 wt.% of BNC (Figure 7A), whereas the tensile strength increased from 18 ± 4 MPa for PMPC/BNC_2 to 69 ± 15 MPa for PMPC/BNC_1 (Figure 7B). In contrast, the elongation at break decreased with the increasing content of BNC from 6.0 ± 1.4% for PMPC/BNC_2 to 2.8 ± 0.3% for PMPC/BNC_1, as shown in Figure 7C. This means that the nanocomposites are more pliable than the stiff BNC nanofibers with an elongation at break of 4.7 ± 1.0%.

The dependence of the membranes' mechanical performance on the amount of BNC is in tune with earlier studies of other BNC-based nanocomposites with polymers of low mechanical properties [14,35,36]. For example, Zhijiang et al. [14] prepared a chitosan/BNC-based hydrogel composite for dye removal, whose Young's modulus increased from 96.5 MP for pure chitosan (dry state) to 244 MPa after the incorporation of BNC nanofibers grafted with carbon nanotubes into the chitosan hydrogel. The same behavior was obtained for the tensile strength and elongation at break [14].

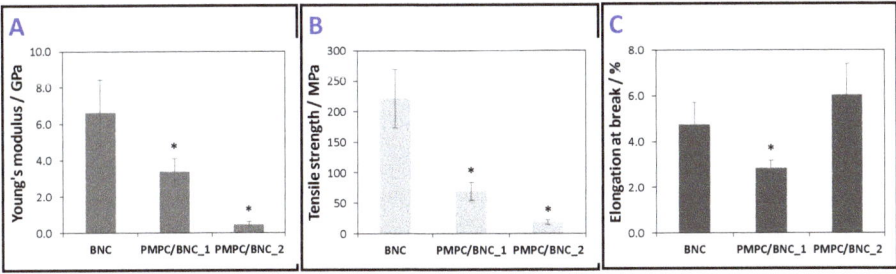

Figure 7. (**A**) Young's modulus, (**B**) tensile strength and (**C**) elongation at break of neat BNC and PMPC/BNC nanocomposites; the error bars correspond to the standard deviations; the asterisk (*) denotes statistically significant differences with respect to the neat BNC ($p < 0.05$).

3.4. In vitro antibacterial activity

Materials with antibacterial activity are relevant for application in multiple fields [37,38] since they can inhibit the growth and simultaneously kill pathogenic bacteria that are harmful to human health [39]. The MPC polymer is known for having antimicrobial and antifouling properties [19–21], which can be a major benefit in reducing/avoiding bacterial growth in contaminated water. This hypothesis was validated by assessing the growth inhibition of gram-positive (*S. aureus*) and gram-negative (*E. coli*) bacteria. *E. coli* was selected for being frequently present in contaminated water, which is a strong indication of recent sewage or fecal contamination. *S. aureus* is not so frequently present in contaminated waters; however, different strains have already been detected in urban wastewater, namely the methicillin-resistant *S. aureus* ST398 [40].

Figure 8 outlines the antibacterial activity of PMPC/BNC nanocomposites and of the neat BNC membrane for comparison purposes. The inoculation of both bacteria in culture media without any sample was used as an experimental control. The neat BNC membrane, along with the experimental control, do not affect the bacterial viability of both *S. aureus* (Figure 8A) and *E. coli* (Figure 8B). This was expected given that BNC is reported not to inhibit the growth of *S. aureus* [41,42], *E. coli* [36,41,42], and other microorganisms such as *Pseudomonas aeruginosa*, *Bacillus subtilis* [42] and *Candida albicans* [43]. In fact, BNC can even be used as a substrate for microbial cell culture [44].

The bacterial killing of *S. aureus* by the two PMPC/BNC nanocomposite membranes is markedly concentration-dependent, as portrayed in Figure 8A. The PMPC/BNC_1 nanocomposite with 54 wt.% of cross-linked PMPC originated a significant reduction ($p < 0.05$) of bacterial concentration relatively to the control, causing a maximum of 2.5–log CFU reduction after 24 h of incubation. The PMPC/BNC_2 with 79 wt.% of cross-linked PMPC reached a higher bacterial inactivation of 4.3–log CFU reduction after 24 h, which indicates that this membrane can be considered an effective antibacterial because according to the American Society of Microbiology (ASM), every new approach has to prove an efficacy of 3–$\log_{10}$ reduction of CFU before being considered antimicrobial or antibacterial [43]. This antibacterial activity is mainly attributed to the trimethylammonium cation that is known for imparting antimicrobial properties [45]. When comparing the activity of the PMPB/BNC membranes with literature, Bertal et al. [46] verified that the triblock copolymer containing PMPC originated an inhibitory zone up to six times greater than the corresponding control against *S. aureus* and a reduction of bacterial growth by 45% compared with the experiments carried out in the absence of PMPC-based copolymer. The authors also claimed that the addition of the copolymer to a 3D-skin model infected with *S. aureus* reduced bacterial recovery by 38% compared to that of controls over 24–48 h [46].

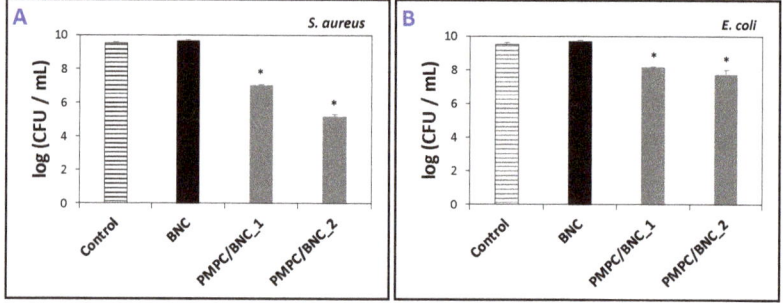

Figure 8. Effect of BNC, PMPC/BNC_1 and PMPC/BNC_2 on the bacterial killing (CFU) of (**A**) *S. aureus* and (**B**) *E. coli* after 24 h of exposure; error bars represent the standard deviation (three independent experiments); the asterisk (*) denotes statistically significant differences with respect to the control and neat BNC ($p < 0.05$).

Regarding the *E. coli* bacteria (Figure 8B), the picture is quite different and both nanocomposites exhibit a lower reduction with 1.3– and 1.8–log CFU reduction for PMPC/BNC_1 and PMPC/BNC_2, respectively. A similar behavior was reported by Fuchs et al. [47] that witnessed no antibacterial activity towards *E. coli* for one MPC copolymer. In fact, this could be expected given that *E. coli* is a gram-negative bacterium whose killing mechanism is more difficult to prevent due to the low permeability of their membranes as discussed previously in detail [48,49].

3.5. Water-uptake and dye removal capacity

Table 2 presents the water-uptake values for BNC and the two PMPC/BNC nanocomposite membranes after immersion in aqueous solutions of 0.01 M HCl (pH 2.1), phosphate buffer saline (pH 7.4) and 0.01 M NaOH (pH 12.0) for 48 h at RT. Overall, the water-uptake vividly increased with the increasing content of cross-linked PMPC. At pH 7.4, it increased from 101 ± 12% for neat BNC up to 639 ± 23% for PMPC/BNC_1 (54 wt.% of PMPC) and 899 ± 44% for PMPC/BNC_2 (79 wt.% of PMPC) (Table 2). In acidic aqueous solutions, PMPC/BNC_1 can absorb 6.3 ± 0.4 g of water *per* g of membrane, while for PMPC/BNC_2 the value is 9.1 ± 0.2 g of water *per* g of membrane. At pH 12, PMPC/BNC_1 absorbs 6.4 ± 0.4 g of water *per* g of membrane, whereas for PMPC/BNC_2 the water-uptake is 9.1 ± 0.3 g of water *per* g of membrane.

Table 2. Water-uptake (water-uptake) of neat BNC and the two PMPC/BNC nanocomposite membranes at different pH media for 48 h at RT.

Membranes	Water-Uptake / %		
	pH 2.1 [a]	pH 7.4 [b]	pH 12.0 [c]
BNC	109 ± 14	101 ± 12	103 ± 10
PMPC/BNC_1	627 ± 38	639 ± 23	640 ± 42
PMPC/BNC_2	911 ± 26	899 ± 44	912 ± 27

[a] Measured after immersion in 0.01 M of HCl aqueous solution; [b] Measured after immersion in phosphate buffer solution; [c] Measured after immersion in 0.01 M of NaOH aqueous solution. All values are the mean of three replicates with the respective standard deviations.

The larger water-uptake of the nanocomposites is correlated with the hydrophilic nature of the phosphorylcholine moiety of the cross-linked PMPC. Additionally, water-uptake is not pH-dependent since there are no significant differences (the means difference is not significant at $\alpha = 0.05$) for the individual membranes under the distinct conditions of acidity or basicity. This can be explained by the unique hydration state of the PMPC chains, where the phosphorylcholine moieties have a hydrophobic hydration layer that do not disturb the hydrogen bonding between the water molecules, as discussed by Ishihara et al. [18]. This is an important characteristic in the water remediation context given that contaminated water can have different pH values. Furthermore, the higher water-uptake of PMPC/BNC_2 is an indication of a higher removal capacity of water-soluble dyes. After 48 h of immersion in aqueous solutions of different pH values, the two nanocomposites were oven dried (at 40 °C) and the final weights demonstrated that the polymer loss ranges between 1%–2%, which emphasizes the insignificant leaching of the cross-linked PMPC from the BNC network.

The removal of two model ionic organic dyes, namely methylene blue (MB) and methyl orange (MO), from water samples at room temperature after 12 h was assessed as a proof-of-concept. While MB is a heterocyclic cationic aromatic compound that is used either as a dye or a drug with for example antimalarial, antidepressant and anxiolytic activity [50], MO is a heterocyclic anionic aromatic compound that is widely used in the textile, pharmaceutical and food industries, and also as an acid-base indicator. Both azo dyes are potentially toxic towards humans and the environment [51].

Figure 9A shows that the PMPC/BNC membranes can indeed retain the model water pollutants as confirmed by the different color of the nanomaterials. This is further corroborated by the data shown in Figure 9B where the dye removal capacity is plotted for each membrane. The pure BNC can remove 0.55 ± 0.12 mg of MB and 0.50 ± 0.06 mg of MO *per* g of membrane. These low removal values were expected, given the lack of binding sites in pure BNC for both cationic and anionic organic dyes. Furthermore, these values are comparable with the dyeability reported by Shim and Kim [52] in their study about the coloration of BNC fabrics with different dyes using *in situ* and *ex situ* methods.

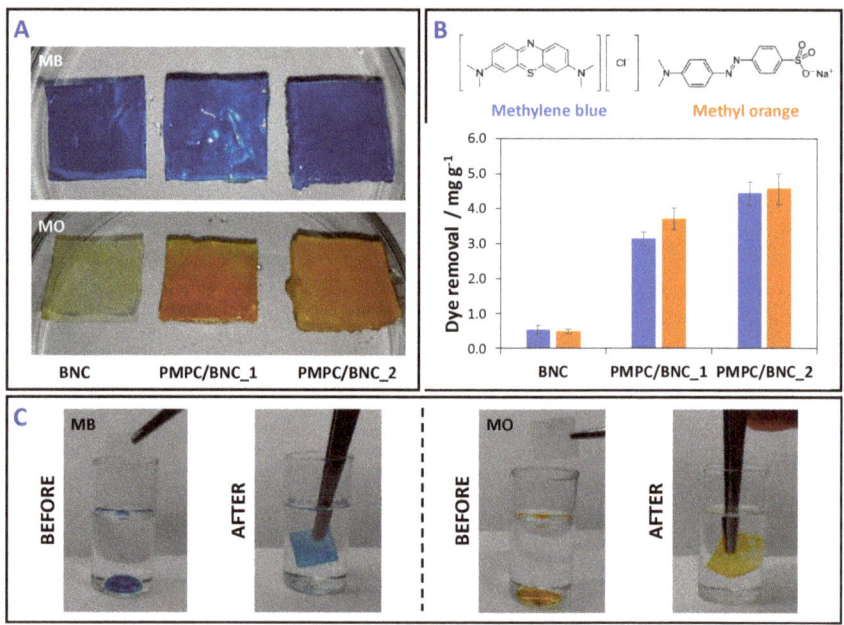

Figure 9. Photographs (**A**) and (**B**) dye removal capacity of BNC, PMPC/BNC_1 and PMPC/BNC_2 after 12 h of immersion in the dye aqueous solution, and (**C**) photographs of the MB and MO aqueous solutions removal from paraffin oil by PMPC/BNC_2 nanocomposite.

Concerning the nanocomposites, PMPC/BNC_1 can remove 3.14 ± 0.19 mg g^{-1} of MB and 3.32 ± 0.31 mg g^{-1} of MO, whereas PMPC/BNC_2 has a removal capacity of 4.44 ± 0.32 mg g^{-1} for MB and 4.56 ± 0.43 mg g^{-1} for MO. Comparing with pure BNC, the dye removal capacity of PMPC/BNC_1 is 5.7 and 7.4 times higher for MB and MO, respectively, while PMPC/BNC_2 removes 8.1 and 9.1 times more MB and MO, respectively, than pure BNC. The higher dye removal capacity of PMPC/BNC_2 is consistent with its higher PMPC content (Table 1). Moreover, the two nanocomposites can remove both cationic and anionic dyes due to the zwitterionic nature of the cross-linked PMPC which can establish electrostatic interactions with either MB or MO model dyes. Worth mentioning is the fact that the PMPC/BNC nanocomposites can easily and quickly remove both MB and MO (25 mg mL^{-1}) from the bottom of a paraffin oil container without the removal of any oil, as exemplified for PMPC/BNC_2 in Figure 9C. This is a good indication of the lack of affinity of the nanocomposites towards the hydrophobic oil and affinity for water or aqueous solutions. A similar behavior was observed for MB (aqueous solution, 100 mg L^{-1}) removal from silicone oil by sulfated-cellulose nanofibrils aerogels [53].

When compared with literature, the dye removal capacity of the PMPC/BNC nanocomposites is lower than that achieved for example with highly carboxylated (COO$^-$) nanocrystalline cellulose with a maximum removal capacity of 101 mg g^{-1} for MB [54], or with the amino-functionalized cellulose nanofibrils-based aerogels with 266 mg g^{-1} for MO [55]. These higher removal capacities are most likely associated with the simultaneous high content of surface binding sites and specific surface area in the first case [54], and the aerogel structure in the second case, which translates into materials with very high porosity and low density [55]. Still, the dye removal values of the PMPC/BNC membranes prepared in the present study are comparable for instance with those achieved with the sulfated-cellulose nanofibrils aerogels that removed *ca.* 5 mg g^{-1} of MB at an adsorbent dosage of 16 mg mL^{-1} [53].

Hence, the adsorbent nanocomposites developed in the present work present a customizable combination of properties, namely antibacterial activity, water-uptake and dye removal capacity, that depend on the amount of the individual components (*i.e.* PMPC and BNC), and that reveal their potential application in the context of water remediation.

4. Conclusions

The combination of the zwitterionic poly(2-methacryloyloxyethyl phosphorylcholine) and the hydrophilic bacterial nanocellulose yielded nanocomposite membranes that are proficient in adsorbing anionic and cationic organic dyes. The optically transparent nanocomposites display high water-uptake capacity in different pH media, thermal stability up to 250 °C, and good mechanical properties (Young's modulus ≥ 430 MPa). Moreover, these zwitterionic membranes inhibited the growth of both Gram-positive (*S. aureus*) and Gram-negative (*E. coli*) pathogenic bacteria with maxima of 4.3– and 1.8–log CFU reduction, respectively, for the nanocomposite composed of 79 wt.% of cross-linked PMPC. Furthermore, their dye removal capacity was demonstrated by a dye adsorption amount of 4.44 ± 0.32 mg g^{-1} of MB and 4.56 ± 0.43 mg g^{-1} of MO for the membrane with the higher content of zwitterionic polymer (*i.e.* 79 wt.% of PMPC). The successful fabrication of these zwitterionic PMPC/BNC nanocomposite membranes with antibacterial activity opens novel avenues for the generation of bio-based adsorbents to address the complex issue of water remediation of anionic and cationic dyes.

Author Contributions: C.V. designed and performed the experiments, analyzed the data and wrote the paper; C.M. and A.A. carried out the antibacterial assays and analyzed the corresponding data; A.J.D.S. and C.S.R.F. contributed to the structural, morphological, thermal and mechanical data interpretation; all authors participated in the critical revision of the paper.

Funding: This work was developed within the scope of the projects CICECO – Aveiro Institute of Materials (UID/CTM/50011/2019) and CESAM (UID/AMB/50017/2019), financed by national funds through the FCT/MEC. The research contract of C. Vilela is funded by national funds (OE), through FCT – Fundação para a Ciência e a Tecnologia, I.P., in the scope of the framework contract foreseen in the numbers 4, 5 and 6 of the article 23, of the Decree-Law 57/2016, of August 29, changed by Law 57/2017, of July 19. FCT is also acknowledge for the research contract under Stimulus of Scientific Employment 2017 to C.S.R. Freire (CEECIND/00464/2017).

Conflicts of Interest: The authors declare no conflict of interest.

References

1. United Nations Transforming Our World: The 2030 Agenda for Sustainable Development. Available online: https://sustainabledevelopment.un.org/post2015/transformingourworld (accessed on 2 March 2019).
2. Li, R.; Zhang, L.; Wang, P. Rational design of nanomaterials for water treatment. *Nanoscale* **2015**, *7*, 17167–17194. [CrossRef] [PubMed]
3. Khan, S.T.; Malik, A. Engineered nanomaterials for water decontamination and purification: From lab to products. *J. Hazard. Mater.* **2019**, *363*, 295–308. [CrossRef] [PubMed]
4. Vilela, C.; Pinto, R.J.B.; Pinto, S.; Marques, P.A.A.P.; Silvestre, A.J.D.; Freire, C.S.R. *Polysaccharide Based Hybrid Materials*, 1st ed.; Springer: Berlin, Germany, 2018; ISBN 978-3-030-00346-3.
5. Corsi, I.; Fiorati, A.; Grassi, G.; Bartolozzi, I.; Daddi, T.; Melone, L.; Punta, C.; Corsi, I.; Fiorati, A.; Grassi, G.; et al. Environmentally sustainable and ecosafe polysaccharide-based materials for water nano-treatment: An eco-design study. *Materials* **2018**, *11*, 1228. [CrossRef]
6. Mahfoudhi, N.; Boufi, S. Nanocellulose as a novel nanostructured adsorbent for environmental remediation: A review. *Cellulose* **2017**, *24*, 1171–1197. [CrossRef]
7. Voisin, H.; Bergström, L.; Liu, P.; Mathew, A. Nanocellulose-based materials for water purification. *Nanomaterials* **2017**, *7*, 57. [CrossRef] [PubMed]
8. Wang, D. A critical review of cellulose-based nanomaterials for water purification in industrial processes. *Cellulose* **2019**, *26*, 687–701. [CrossRef]

9. Jacek, P.; Dourado, F.; Gama, M.; Bielecki, S. Molecular aspects of bacterial nanocellulose biosynthesis. *Microb. Biotechnol.* **2019**. [CrossRef] [PubMed]
10. Figueiredo, A.R.P.; Vilela, C.; Neto, C.P.; Silvestre, A.J.D.; Freire, C.S.R. Bacterial cellulose-based nanocomposites: Roadmap for innovative materials. In *Nanocellulose Polymer Composites*; Thakur, V.K., Ed.; Scrivener Publishing LLC: Salem, MA, USA, 2015; pp. 17–64.
11. Vilela, C.; Pinto, R.J.B.; Figueiredo, A.R.P.; Neto, C.P.; Silvestre, A.J.D.; Freire, C.S.R. Development and applications of cellulose nanofibers based polymer composites. In *Advanced Composite Materials: Properties and Applications*; Bafekrpour, E., Ed.; De Gruyter Open: Berlin, Germany, 2017; pp. 1–65.
12. Torres, F.G.; Arroyo, J.J.; Troncoso, O.P. Bacterial cellulose nanocomposites: An all-nano type of material. *Mater. Sci. Eng. C* **2019**, *98*, 1277–1293. [CrossRef]
13. Chen, S.; Huang, Y. Bacterial cellulose nanofibers decorated with phthalocyanine: Preparation, characterization and dye removal performance. *Mater. Lett.* **2015**, *142*, 235–237. [CrossRef]
14. Zhijiang, C.; Ping, X.; Cong, Z.; Tingting, Z.; Jie, G.; Kongyin, Z. Preparation and characterization of a bi-layered nano-filtration membrane from a chitosan hydrogel and bacterial cellulose nanofiber for dye removal. *Cellulose* **2018**, *25*, 5123–5137. [CrossRef]
15. Sai, H.; Fu, R.; Xing, L.; Xiang, J.; Li, Z.; Li, F.; Zhang, T. Surface modification of bacterial cellulose aerogels' web-like skeleton for oil/water separation. *ACS Appl. Mater. Interfaces* **2015**, *7*, 7373–7381. [CrossRef]
16. Lu, M.; Guan, X.-H.; Xu, X.-H.; Wei, D.-Z. Characteristic and mechanism of Cr(VI) adsorption by ammonium sulfamate-bacterial cellulose in aqueous solutions. *Chin. Chem. Lett.* **2013**, *24*, 253–256. [CrossRef]
17. Huang, X.; Zhan, X.; Wen, C.; Xu, F.; Luo, L. Amino-functionalized magnetic bacterial cellulose/activated carbon composite for Pb^{2+} and methyl orange sorption from aqueous solution. *J. Mater. Sci. Technol.* **2018**, *34*, 855–863. [CrossRef]
18. Ishihara, K.; Mu, M.; Konno, T.; Inoue, Y.; Fukazawa, K. The unique hydration state of poly(2-methacryloyloxyethyl phosphorylcholine). *J. Biomater. Sci. Polym. Ed.* **2017**, *28*, 884–899. [CrossRef]
19. Goda, T.; Ishihara, K.; Miyahara, Y. Critical update on 2-methacryloyloxyethyl phosphorylcholine (MPC) polymer science. *J. Appl. Polym. Sci.* **2015**, *132*, 41766. [CrossRef]
20. Fujiwara, N.; Yumoto, H.; Miyamoto, K.; Hirota, K.; Nakae, H.; Tanaka, S.; Murakami, K.; Kudo, Y.; Ozaki, K.; Miyake, Y. 2-Methacryloyloxyethyl phosphorylcholine (MPC)-polymer suppresses an increase of oral bacteria: A single-blind, crossover clinical trial. *Clin. Oral Investig.* **2019**, *23*, 739–746. [CrossRef] [PubMed]
21. Kwon, J.-S.; Lee, M.-J.; Kim, J.-Y.; Kim, D.; Ryu, J.-H.; Jang, S.; Kim, K.-M.; Hwang, C.-J.; Choi, S.-H. Novel anti-biofouling bioactive calcium silicate-based cement containing 2-methacryloyloxyethyl phosphorylcholine. *PLoS ONE* **2019**, *14*, e0211007. [CrossRef] [PubMed]
22. Trovatti, E.; Serafim, L.S.; Freire, C.S.R.; Silvestre, A.J.D.; Neto, C.P. Gluconacetobacter sacchari: An efficient bacterial cellulose cell-factory. *Carbohydr. Polym.* **2011**, *86*, 1417–1420. [CrossRef]
23. Vilela, C.; Sousa, N.; Pinto, R.J.B.; Silvestre, A.J.D.; Figueiredo, F.M.L.; Freire, C.S.R. Exploiting poly(ionic liquids) and nanocellulose for the development of bio-based anion-exchange membranes. *Biomass Bioenergy* **2017**, *100*, 116–125. [CrossRef]
24. Saïdi, L.; Vilela, C.; Oliveira, H.; Silvestre, A.J.D.; Freire, C.S.R. Poly(N-methacryloyl glycine)/nanocellulose composites as pH-sensitive systems for controlled release of diclofenac. *Carbohydr. Polym.* **2017**, *169*, 357–365. [CrossRef]
25. Barud, H.S.; Ribeiro, S.J.L.; Carone, C.L.P.; Ligabue, R.; Einloft, S.; Queiroz, P.V.S.; Borges, A.P.B.; Jahno, V.D. Optically transparent membrane based on bacterial cellulose/polycaprolactone. *Polímeros* **2013**, *23*, 135–138. [CrossRef]
26. Pleumphon, C.; Thiangtham, S.; Pechyen, C.; Manuspiya, H.; Ummartyotin, S. Development of conductive bacterial cellulose composites: An approach to bio-based substrates for solar cells. *J. Biobased Mater. Bioenergy* **2017**, *11*, 321–329. [CrossRef]
27. Foster, E.J.; Moon, R.J.; Agarwal, U.P.; Bortner, M.J.; Bras, J.; Camarero-Espinosa, S.; Chan, K.J.; Clift, M.J.D.; Cranston, E.D.; Eichhorn, S.J.; et al. Current characterization methods for cellulose nanomaterials. *Chem. Soc. Rev.* **2018**, *47*, 2609–2679. [CrossRef] [PubMed]
28. Bellamy, L.J. *The Infrared Spectra of Complex Molecules*, 3rd ed.; Chapman and Hall, Ltd.: London, UK, 1975; ISBN 0412138506.
29. Xu, S.; Ye, Z.; Wu, P. Biomimetic controlling of $CaCO_3$ and $BaCO_3$ superstructures by zwitterionic polymer. *ACS Sustain. Chem. Eng.* **2015**, *3*, 1810–1818. [CrossRef]

30. Ishihara, K.; Ueda, T.; Nakabayashi, N. Preparation of phospholipid polymers and their properties as polymer hydrogel membranes. *Polym. J.* **1990**, *22*, 355–360. [CrossRef]
31. Vilela, C.; Martins, A.P.C.; Sousa, N.; Silvestre, A.J.D.; Figueiredo, F.M.L.; Freire, C.S.R. Poly(bis[2-(methacryloyloxy)ethyl] phosphate)/bacterial cellulose nanocomposites: Preparation, characterization and application as polymer electrolyte membranes. *Appl. Sci.* **2018**, *8*, 1145. [CrossRef]
32. Gadim, T.D.O.; Figueiredo, A.G.P.R.; Rosero-Navarro, N.C.; Vilela, C.; Gamelas, J.A.F.; Barros-Timmons, A.; Neto, C.P.; Silvestre, A.J.D.; Freire, C.S.R.; Figueiredo, F.M.L. Nanostructured bacterial cellulose-poly(4-styrene sulfonic acid) composite membranes with high storage modulus and protonic conductivity. *ACS Appl. Mater. Interfaces* **2014**, *6*, 7864–7875. [CrossRef]
33. Wang, S.; Liu, Q.; Luo, Z.; Wen, L.; Cen, K. Mechanism study on cellulose pyrolysis using thermogravimetric analysis coupled with infrared spectroscopy. *Front. Energy Power Eng. China* **2007**, *1*, 413–419. [CrossRef]
34. Faria, M.; Vilela, C.; Mohammadkazemi, F.; Silvestre, A.J.D.; Freire, C.S.R.; Cordeiro, N. Poly(glycidyl methacrylate)/ bacterial cellulose nanocomposites: Preparation, characterization and post-modification. *Int. J. Biol. Macromol.* **2019**, *127*, 618–627. [CrossRef]
35. Vilela, C.; Gadim, T.D.O.; Silvestre, A.J.D.; Freire, C.S.R.; Figueiredo, F.M.L. Nanocellulose/ poly(methacryloyloxyethyl phosphate) composites as proton separator materials. *Cellulose* **2016**, *23*, 3677–3689. [CrossRef]
36. Figueiredo, A.R.P.; Figueiredo, A.G.P.R.; Silva, N.H.C.S.; Barros-Timmons, A.; Almeida, A.; Silvestre, A.J.D.; Freire, C.S.R. Antimicrobial bacterial cellulose nanocomposites prepared by in situ polymerization of 2-aminoethyl methacrylate. *Carbohydr. Polym.* **2015**, *123*, 443–453. [CrossRef]
37. Vilela, C.; Kurek, M.; Hayouka, Z.; Röcker, B.; Yildirim, S.; Antunes, M.D.C.; Nilsen-Nygaard, J.; Pettersen, M.K.; Freire, C.S.R. A concise guide to active agents for active food packaging. *Trends Food Sci. Technol.* **2018**, *80*, 212–222. [CrossRef]
38. Bui, V.; Park, D.; Lee, Y.-C. Chitosan combined with ZnO, TiO$_2$ and Ag nanoparticles for antimicrobial wound healing applications: A mini review of the research trends. *Polymers* **2017**, *9*, 21. [CrossRef]
39. Konai, M.M.; Bhattacharjee, B.; Ghosh, S.; Haldar, J. Recent progress in polymer research to tackle infections and antimicrobial resistance. *Biomacromolecules* **2018**, *19*, 1888–1917. [CrossRef]
40. Gómez, P.; Lozano, C.; Benito, D.; Estepa, V.; Tenorio, C.; Zarazaga, M.; Torres, C. Characterization of staphylococci in urban wastewater treatment plants in Spain, with detection of methicillin resistant Staphylococcus aureus ST398. *Environ. Pollut.* **2016**, *212*, 71–76. [CrossRef]
41. Padrão, J.; Gonçalves, S.; Silva, J.P.; Sencadas, V.; Lanceros-Méndez, S.; Pinheiro, A.C.; Vicente, A.A.; Rodrigues, L.R.; Dourado, F. Bacterial cellulose-lactoferrin as an antimicrobial edible packaging. *Food Hydrocoll.* **2016**, *58*, 126–140. [CrossRef]
42. Shao, W.; Wang, S.; Wu, J.; Huang, M.; Liu, H.; Min, H. Synthesis and antimicrobial activity of copper nanoparticle loaded regenerated bacterial cellulose membranes. *RSC Adv.* **2016**, *6*, 65879–65884. [CrossRef]
43. Vilela, C.; Oliveira, H.; Almeida, A.; Silvestre, A.J.D.; Freire, C.S.R. Nanocellulose-based antifungal nanocomposites against the polymorphic fungus Candida albicans. *Carbohydr. Polym.* **2019**. [CrossRef]
44. Yin, N.; Santos, T.M.A.; Auer, G.K.; Crooks, J.A.; Oliver, P.M.; Weibel, D.B. Bacterial cellulose as a substrate for microbial cell culture. *Appl. Environ. Microbiol.* **2014**, *80*, 1926–1932. [CrossRef]
45. Li, J.; Cha, R.; Mou, K.; Zhao, X.; Long, K.; Luo, H.; Zhou, F.; Jiang, X. Nanocellulose-based antibacterial materials. *Adv. Healthc. Mater.* **2018**, *7*, 1800334. [CrossRef]
46. Bertal, K.; Shepherd, J.; Douglas, C.W.I.; Madsen, J.; Morse, A.; Edmondson, S.; Armes, S.P.; Lewis, A.; MacNeil, S. Antimicrobial activity of novel biocompatible wound dressings based on triblock copolymer hydrogels. *J. Mater. Sci.* **2009**, *44*, 6233–6246. [CrossRef]
47. Fuchs, A.V.; Ritz, S.; Pütz, S.; Mailänder, V.; Landfester, K.; Ziener, U. Bioinspired phosphorylcholine containing polymer films with silver nanoparticles combining antifouling and antibacterial properties. *Biomater. Sci.* **2013**, *1*, 470–477. [CrossRef]
48. Band, V.I.; Weiss, D.S. Mechanisms of antimicrobial peptide resistance in gram-negative bacteria. *Antibiotics* **2014**, *4*, 18–41. [CrossRef] [PubMed]
49. Masi, M.; Réfregiers, M.; Pos, K.M.; Pagès, J.M. Mechanisms of envelope permeability and antibiotic influx and efflux in Gram-negative bacteria. *Nat. Microbiol.* **2017**, *2*, 17001. [CrossRef]
50. Delport, A.; Harvey, B.H.; Petzer, A.; Petzer, J.P.; Brain Dis, M. Methylene blue and its analogues as antidepressant compounds. *Metab. Brain Dis.* **2017**, *32*, 1357–1382. [CrossRef] [PubMed]

51. Forgacs, E.; Cserháti, T.; Oros, G. Removal of synthetic dyes from wastewaters: A review. *Environ. Int.* **2004**, *30*, 953–971. [CrossRef] [PubMed]
52. Shim, E.; Kim, H.R. Coloration of bacterial cellulose using in situ and ex situ methods. *Text. Res. J.* **2018**, *89*, 1297–1310. [CrossRef]
53. Wang, D.C.; Yu, H.Y.; Fan, X.; Gu, J.; Ye, S.; Yao, J.; Ni, Q.Q. High aspect ratio carboxylated cellulose nanofibers cross-linked to robust aerogels for superabsorption-flocculants: Paving way from nanoscale to macroscale. *ACS Appl. Mater. Interfaces* **2018**, *10*, 20755–20766. [CrossRef]
54. He, X.; Male, K.B.; Nesterenko, P.N.; Brabazon, D.; Paull, B.; Luong, J.H.T. Adsorption and desorption of methylene blue on porous carbon monoliths and nanocrystalline cellulose. *ACS Appl. Mater. Interfaces* **2013**, *5*, 8796–8804. [CrossRef]
55. Tang, J.; Song, Y.; Zhao, F.; Spinney, S.; Bernardes, J.S.; Tam, K.C. Compressible cellulose nanofibril (CNF) based aerogels produced via a bio-inspired strategy for heavy metal ion and dye removal. *Carbohydr. Polym.* **2019**, *208*, 404–412. [CrossRef]

© 2019 by the authors. Licensee MDPI, Basel, Switzerland. This article is an open access article distributed under the terms and conditions of the Creative Commons Attribution (CC BY) license (http://creativecommons.org/licenses/by/4.0/).

Article

One-Minute Synthesis of Size-Controlled Fucoidan-Gold Nanosystems: Antitumoral Activity and Dark Field Imaging

Ricardo J. B. Pinto [1,*], Daniela Bispo [1], Carla Vilela [1], Alexandre M. P. Botas [2], Rute A. S. Ferreira [2], Ana C. Menezes [3], Fábio Campos [3], Helena Oliveira [3], Maria H. Abreu [4], Sónia A. O. Santos [1] and Carmen S. R. Freire [1,*]

[1] Department of Chemistry, CICECO—Aveiro Institute of Materials, University of Aveiro, 3810-193 Aveiro, Portugal; d.bispo@ua.pt (D.B.); cvilela@ua.pt (C.V.); santos.sonia@ua.pt (S.A.O.S.)
[2] Phantom-G, Department of Physics, CICECO—Aveiro Institute of Materials, University of Aveiro, 3810-193 Aveiro, Portugal; a.botas@ua.pt (A.M.P.B.); rferreira@ua.pt (R.A.S.F.)
[3] Department of Biology & CESAM, University of Aveiro, 3810-193 Aveiro, Portugal; catarinamenezes@msn.com (A.C.M.); f.m.c@ua.pt (F.C.); holiveira@ua.pt (H.O.)
[4] ALGAplus—Prod. e Comerc. De Algas e Seus Derivados, Lda., 3830-196 Ílhavo, Portugal; helena.abreu@algaplus.pt
* Correspondence: r.pinto@ua.pt (R.J.B.P.); cfreire@ua.pt (C.S.R.F.)

Received: 6 February 2020; Accepted: 24 February 2020; Published: 28 February 2020

Abstract: Gold nanoparticles (AuNPs) are one of the most studied nanosystems with great potential for biomedical applications, including cancer therapy. Although some gold-based systems have been described, the use of green and faster methods that allow the control of their properties is of prime importance. Thus, the present study reports a one-minute microwave-assisted synthesis of fucoidan-coated AuNPs with controllable size and high antitumoral activity. The NPs were synthesized using a fucoidan-enriched fraction extracted from *Fucus vesiculosus*, as the reducing and capping agent. The ensuing monodispersed and spherical NPs exhibit tiny diameters between 5.8 and 13.4 nm for concentrations of fucoidan between 0.5 and 0.05% (w/v), respectively, as excellent colloidal stability in distinct solutions and culture media. Furthermore, the NPs present antitumoral activity against three human tumor cell lines (MNT-1, HepG2, and MG-63), and flow cytometry in combination with dark-field imaging confirmed the cellular uptake of NPs by MG-63 cell line.

Keywords: gold nanoparticles; fucoidan; microwave irradiation; antitumoral activity; darkfield imaging

---

1. Introduction

Cancer, i.e., the abnormal growth and proliferation of cells, is one of the leading causes of mortality and morbidity worldwide. According to the World Health Organization (WHO), cancer was responsible for 9.9 million deaths in 2018 [1], and the number of cases is anticipated to increase by about 70% over the next two decades. Each cancer type involves a specific treatment procedure that embraces one or more modalities, namely surgery (to remove the tumor), radiotherapy, or chemotherapy. Despite the high cure rates observed when cancer is detected early and if appropriated treatment is provided, most of the presently employed therapies, particularly conventional chemotherapy, are associated with severe side effects, including hair loss, nausea and vomiting, pain, anemia, fertility issues, edema, among many others [2], that strongly affect the patient's quality of life. In recent decades, the design of more effective alternatives that allows a targeting action, with almost no impact on healthy tissues or organs, has received considerable attention [3]. Of these, nanosystems that combine a therapeutic effect and imaging properties, or that promote intertwined diagnosis and therapy, the so-called nanotheranostics, has opened new avenues for cancer-conquering [4–6].

Metal nanoparticles (NPs) are a class of nanomaterials with a panoply of biomedical and therapeutic applications, including cancer therapy and imaging of tumors [7,8]. In particular, gold (Au) nanoparticles owing to their unique physical and optical properties have attracted enormous interest in this realm during the past decades [9–12]. The colloidal stability of NPs in biological environments and their interactions with cells is strongly influenced by their surface properties, and thus distinct coating strategies of NPs, using small molecules, polymers, or lipids, have been described [13]. Moreover, this methodology could also target the reduction of the toxicity of NPs and improvement of biological functionalities, typically associated with the use of active biomolecules [13]. The synthesis of AuNPs using biomacromolecules, including alginate, starch, cellulose, chitosan, gelatin, collagen and fucoidan, among others, as reducing and stabilizing agents is a well-documented strategy to achieve these goals [14] and, at the same time, overcome the environmental effects of the conventional methodologies that commonly involve the use of harmful reducing agents.

Fucoidan is a natural occurring sulfated marine polysaccharide extracted from brown seaweeds that presents various biological properties, including antiangiogenic, antitumoral, and anti-inflammatory properties [15,16], and because of that has been widely investigated for the development of nanomaterials for biomedical applications [17]. However, only a few number of papers reported the combination of fucoidan and AuNPs for cancer treatment. One of the first studies in this topic involved the synthesis of AuNPs (44 nm average size) using sodium borohydride as the reducing agent and a fucoidan-mimetic glycopolymer as the capping agent [18]. The obtained AuNPs displayed excellent colloidal stability and selective cytotoxicity to human colon cancer cell line (HCT116). Afterward, Manivasagan et al. [19] produced biocompatible AuNPs (82 nm average size) by using naturally occurring fucoidan as the reducing and capping agent, avoiding the use of harmful reducing agents. This study also demonstrated the applicability of these NPs as a carrier for doxorubicin (DOX) and photoacoustic imaging of breast cancer tumors. In a follow-up study, this research team explored similar fucoidan-AuNPs for dual-chemo-photothermal treatment of eye tumors [20]. In another study, size-controlled fucoidan-AuNPs (15-80 nm) were produced by varying the concentration of fucoidan during the synthesis step [21]. These NPs showed anticancer effect against human oral squamous cell carcinoma (HSC3), and its surface modification and conjugation with DOX also improved their effect.

These studies clearly demonstrate the prospective of the partnership between AuNPs and fucoidan in cancer treatment. However, to achieve high antitumoral activities, viz., less than 20% cell viability, the conjugation with other chemotherapeutics (DOX), or the use of high dosages (up to 50 μg mL^{-1} of NPs) was typically required. Moreover, some methodologies are somewhat time-consuming and laborious, e.g., up to 2 h for the synthesis of the AuNPs and 24 h for the conjugation with DOX. Additionally, the number of investigated tumor cell lines is limited, and the imaging properties of fucoidan-AuNPs have been only marginally explored. Thus, some essential traits still need to be tackled envisioning their scale-up production and broad application, viz., the establishment of fast and straightforward procedures for the synthesis of fucoidan-Au nanosystems with controllable size and morphology and improved antitumoral activity against different tumor cell lines.

As a developing heating tool, microwave irradiation has been shown to considerably reduce the reaction times and provide a uniform bulk heating that allows the synthesis of various nanomaterials, including AuNPs [22,23], with defined structures and narrow size distributions. However, to the best of our knowledge, this methodology has never been sightseen as a simple, time-saving approach to fabricate fucoidan-AuNPs for application in cancer therapy.

In this line, in the present study, we report for the first time a one-minute microwave-assisted synthesis of fucoidan-coated AuNPs with controllable size and high antitumoral activity. This is a pioneering achievement with respect to previous methodologies to produce Au-fucoidan NPs. The fucoidan-AuNPs were synthesized, using a fucoidan-enriched fraction extracted from *F. vesiculosus*, and characterized in terms of structure, colloidal stability, antitumoral activity against different cell lines (MNT-1, HepG2, and MG-63), and cellular uptake by flow cytometry and dark field imaging of

NPs. The antitumoral activity of Au-fucoidan NPs against these cell lines and their imaging properties by dark-field is also reported here for the first time.

2. Materials and Methods

2.1. Materials

Fucus vesiculosus was collected (January 2014) in Mindelo beach (41°18′38′′N, 8°,43′42′′W), Portugal. The biomass was washed with water to remove salts, epiphytes, and/or microorganisms and dried at 25 °C until reaching a total moisture content of 12–14%. Algae samples were transformed into flakes (1–2 mm) with a knife mill (Retsch SM100, Haan, Germany), packed and stored in airtight bags at the ALGAplus warehouse. The milled algae samples were then washed with a solvent mixture (1 g per 20 mL) of chloroform and methanol (2:1 v/v) under stirring for 20 min, centrifuged at 2500 rpm during 20 min, and dried at 40 °C in a vacuum drying oven.

Gold(III) chloride trihydrate (≥99.9% trace metals basis), dimethyl sulfoxide (DMSO), paraformaldehyde, and Triton X-100 were purchased from Sigma-Aldrich (St. Louis, MO, USA). Phosphate buffer saline (PBS, pH 7.4), Dulbecco's modified Eagle's medium (DMEM), fetal bovine serum (FBS), L-glutamine, penicillin, streptomycin and amphotericin B were supplied by Gibco® (Life Technologies, Grand Island, NY, USA), 3-(4,5-dimethylthiazol-2-yl)-2,5-diphenyltetrazolium bromide (MTT, 98%) was purchased from Sigma-Aldrich, and 4′,6-diamidino-2-phenylindole (DAPI)-containing Vectashield mounting medium was acquired from Vector Labs. Ultra-purified water (Type 1, 18.2 MΩ·cm at 25 °C) was obtained by a Simplicity® Water Purification System (Merck, Darmstadt, Germany). Human osteosarcoma cell line MG-63 was a kind gift by INEB, University of Porto, Portugal. The HepG2 cell line, a liver hepatocellular carcinoma cell line, was obtained from the European Collection of Authenticated Cell Cultures (ECACC, Salisbury, UK) and supplied by Sigma-Aldrich. MNT-1 cells were kindly provided by Doctor Manuela Gaspar (iMed.ULisboa, Lisbon, Portugal).

2.2. Microwave-Assisted Extraction (MAE) of Fucoidan from F. Vesiculosus

The MAE of fucoidan-rich fraction from *F. vesiculosus* followed the methodology described by Rodriguez-Jasso et al. [24]. About 0.4 g of macroalgae sample was suspended in water, with a solid-liquid ratio of 1:25 (w/v). The extraction was performed in a Monowave 300 (Anton Paar, Graz, Austria) equipment, at 172 °C, for 1 min. The samples were cooled on ice and then centrifuged at 4000 rpm during 5 min). The aqueous extract was mixed with a 1% (w/v) $CaCl_2$ aqueous solution, in a solid-liquid ratio of 1:1 (v/v), and maintained overnight at 4 °C, to precipitate alginate. This was separated by filtration, and ethanol was added to the filtrate (1:2, v/v) and maintained at 4 °C for 8 h. The fucoidan-rich fraction (MWF) was obtained after centrifugation (4000 rpm, 5 min) and dried at room temperature.

2.3. Microwave-Assisted Synthesis of Fucoidan-AuNPs

The microwave (MW)-assisted synthesis of fucoidan-AuNPs was carried out on a Monowave 300 equipment (Anton Paar, firmware version 2.0). Total of 145 µL of $HAuCl_4·3H_2O$ solution (17.2 mM) was added to microwave vials with 5 mL of the three distinct concentrations of the fucoidan-rich fraction (0.5, 0.1, and 0.05% w/v). The mixtures were heated at 120 °C for 1 min. After the reaction, the obtained fucoidan-AuNPs were centrifuged for 1 h at 15,000 rpm and 4 °C, sonicated and washed three times with ultra-purified water, and finally stored at 4 °C until usage.

2.4. Structural Characterization of the Fucoidan-Enriched Fraction and Fucoidan-AuNPs

The sulfate content of the fucoidan-enriched fraction was determined by elemental analysis. The fucoidan-rich fraction was grounded with a mortar and analyzed (about 2–3 mg) in a Leco TruSpec 630-200-200 CHNS elemental analyzer (LECO Corporation, St. Joseph, MI, USA), in order to assess the carbon (C), hydrogen (H), nitrogen (N), and sulfur (S) contents. The sulfate content of the sample was calculated by the following equation: sulfate content (%) = 3.22 × S (%), where S (%) is the S content, as proposed by several authors [25,26].

Total sugars were determined by the phenol-sulfuric acid method, following the methodology described by DuBois et al. [27] where D(+)-glucose was used as standard. An aqueous solution of the fucoidan-rich fraction (0.05% w/v) was prepared and diluted as necessary. The UV absorbance measurements were performed in a Shimadzu UV-1800 spectrophotometer (Shimadzu Corp., Kyoto, Japan) at λ = 490 nm. Triplicate measurements were carried out.

Optical spectra of the fucoidan-AuNPs were recorded by a Thermo Scientific Evolution 220 spectrophotometer (Thermo Fisher Scientific, Waltham, MA, USA) using 100 scans min^{-1} with a bandwidth of 2 nm and an integration time of 0.3 s.

The Fourier transform infrared (FTIR) spectra of fucoidan-enriched fraction extracted from *F. vesiculosus* and fucoidan-Au colloidal in the form of KBr pellet were collected by a Mattson 7000 spectrometer using 256 scans at a resolution of 4 cm^{-1} and with a signal gain of 1.

Transmission electron microscopy (TEM) images were obtained by a Hitachi H-9000 (Hitachi High-Technologies Corporation, Tokyo, Japan) operating at 300 kV. Scanning transmission electron microscopy (STEM) images were acquired by a field-emission gun (FEG) SEM Hitachi SU70 microscope operated at 15 kV. Samples for microscopy analysis were prepared by placing a washed colloid drop directly onto a carbon-coated copper grid and then allowing the solvent to evaporate. The average diameter of the NPs was determined by measuring over 100 NPs for each STEM image with the Fiji image processing software.

2.5. Colloidal Stability of Fucoidan-AuNPs

The colloidal stability of the fucoidan-AuNPs was evaluated in five different mediums, namely ultra-purified water, HCl (0.01 M, pH 2.1), NaOH (0.01 M, pH 12.0), PBS (pH 7.4), and DMEM. Typically, 100 µL of the fucoidan-AuNPs 0.1% (w/v) was added to the vials already with 2.9 mL of the distinct mediums. The colloidal suspensions were placed under mechanical stirring during 48 h at room temperature, and the UV-Vis spectra at specific times (0, 6, 24, and 48 h) were recorded. All assays were performed in triplicate. After 48 h, the NPs in each medium were centrifuged for 15 min at 15,000 rpm (4 °C), sonicated and washed three times with ultra-purified water and analyzed by STEM as previously described.

2.6. Cell Culture

The cell lines were cultured in DMEM supplemented with 10% fetal bovine serum, 2 mM L-glutamine, 100 U mL^{-1} penicillin, 100 µg mL^{-1} streptomycin, and 2.5 µg mL^{-1} amphotericin B. Cells were incubated in a humidified atmosphere of 5% CO_2 at 37 °C. Sub confluent cells were trypsinized with trypsin-EDTA (0.25% trypsin, 1 mM EDTA) when monolayers reached 70% confluence.

2.7. Cytotoxicity Evaluation

Cell viability was determined by the colorimetric MTT assay, which measures the formation of purple formazan in viable cells [28]. Cells were seeded in 96-well plates and after 24 h, medium was replaced with fresh medium containing fucoidan extract (0, 0.25, 0.5, 1.0, 2.5, and 5 mg mL^{-1}) or fucoidan-AuNPs (0, 2.5, 5.0, 10.0, 15.0, and 26.0 µg mL^{-1}). Cell viability was measured after 24, 48, and 72 h. After that, 50 µL of MTT reagent (1 mg mL^{-1}) in PBS was added to each well and incubated for 4 h at 37 °C, and 5% CO_2. The medium was then removed, and 150 µL of DMSO was added to each well for crystals solubilization. The optical density of the reduced MTT was measured at 570 nm in a microtiter plate reader (Synergy HT Multi-Mode, BioTeK instruments, Winooski, VT, USA), and the cell metabolic activity (MA, a usual marker for cell viability) was calculated as MA (%) = ((Abs $_{sample}$−Abs $_{DMSO}$)/(Abs $_{control}$−Abs $_{DMSO}$)) × 100. Three independent assays were performed with at least three technical replicates each and the results compared with the control (incubated with culture medium).

From the MTT results, the concentrations of 5 and 12 µg mL^{-1} of fucoidan-AuNPs were selected for the following assays.

2.8. Uptake Potential by Flow Cytometry

The uptake potential of fucoidan-AuNPs by MG-63 cells was assessed by flow cytometry (FCM), as previously described by Suzuki et al. [29] and Bastos et al. [30]. Briefly, cells were seeded in 6-well plates, and after fucoidan-AuNPs exposure for 24 h, they were trypsinized, collected to FCM tubes, and analyzed by FCM. Two parameters, namely forward scatter (FS), which gives information on the particle's size, and side scatter (SS), which provides information on the complexity of particles, were measured in an Attune® Acoustic Focusing Cytometer (Thermo Scientific, Waltham, MA, USA) equipped with a 488 nm laser. For each sample, 5000–20,000 cells were analyzed at a flow rate of about 300 cells s^{-1}.

For MTT assay and cellular uptake by flow cytometry, the statistical significance between control and exposed cells was performed by one-way ANOVA, followed by Dunnet and Dunn's method (as parametric and non-parametric test, respectively), using Sigma Plot 12.5 software (Systat Software Inc.).

2.9. Dark Field Imaging

MG-63 cells were grown on glass coverslips and cultured in the presence of 12 µg mL^{-1} fucoidan-AuNPs dispersed in culture medium for 24 h. Cells were fixed with a 4% paraformaldehyde in PBS for 10 min, permeabilized with a 0.1% Triton X-100/PBS solution. Following washes with PBS and deionized water, coverslips were mounted onto the glass slides with DAPI-containing Vectashield mounting medium.

The microscopic images were recorded using an Olympus BX51 microscope (50× objective) (Olympus, Tokyo, Japan) equipped with a digital CCD camera (Retiga 4000R, QImaging) used to capture the microphotographs. The dark field images were acquired under white light illumination by replacing the standard microscope condenser by the CytoViva enhanced dark field illumination system (CytoViva, Auburn, AL, USA). For the images under white light illumination and UV irradiation, a DC regulated illuminator (DC-950, Fiber-Lite) and a LED light (LLS-365, Ocean Optics, emission at 365 ± 25 nm) were used, respectively.

The hyperspectral images were recorded with a hyperspectral imaging system from CytoViva, accouped to the Olympus BX51 microscope, that includes a digital camera (Pixelfly USB, PCO) coupled to a spectrograph (V10E 2/3″, Specim, 30 µm slit, nominal spectral range of 400–1000 nm and nominal spectral resolution of 2.73 nm). Each pixel field-of-view on the hyperspectral images corresponds to 258 × 258 nm^2 on the samples' plane. The hyperspectral scanning is vertical, and each image results from 696 lines, using an exposure time of 3 s for each line. All the hyperspectral data were acquired and analyzed using ENVI 4.8 software, and the spectra were corrected using the tool Calibration and Correction of the ENVI 4.8 software.

3. Results and Discussion

A fucoidan-rich fraction from *F. vesiculosus* was used as the reducing and capping agent in the green synthesis of antitumoral fucoidan-AuNPs for application in cancer therapy (Figure 1). MW technology was used for both the extraction of fucoidan from *F. vesiculosus* and the synthesis of the AuNPs, pursuing the establishment of a timesaving methodology to produce fucoidan-Au nanosystems with controllable size, morphology, and high antitumoral effect, as will be discussed in the following paragraphs.

The extraction yield obtained for the fucoidan-rich fraction was 5.2 ± 0.8% that is in good agreement with the published data for this algae specie under similar MAE conditions [31]. However, the degree of sulfation (5.42%) and total sugars content (13.7 ± 0.6%) are lower than those previously reported [31]. These differences are certainly associated with the natural variability of seaweed biomass [32]. The presence of fucoidan was further confirmed by FTIR analysis. The FTIR spectrum of the fucoidan-rich fraction (Figure 2) displays the typical absorption bands of fucoidans [31], namely a band at around 1260 cm^{-1} (asymmetric stretching of the O=S=O groups of sulfate esters, with the contribution of C–OH, C–C and C–O vibrations). Moreover, a band at 840 cm^{-1} (C–O–S bending associated with the axial substitution at C-4 position) and a shoulder at around 820 cm^{-1} (C–O–S bending associated with substitution at C-2 and C-3 equatorial positions of fucopyranosyl moieties) are also present [33]. The higher intensity of the band at 820 cm^{-1} suggests that the extracted fucoidan is mainly characterized by repeated units of disaccharides primarily composed of fucose residues with -OSO$_3^-$ at C-2 and C-3 positions and with a single -OSO$_3^-$ at C-2 position (Figure 2).

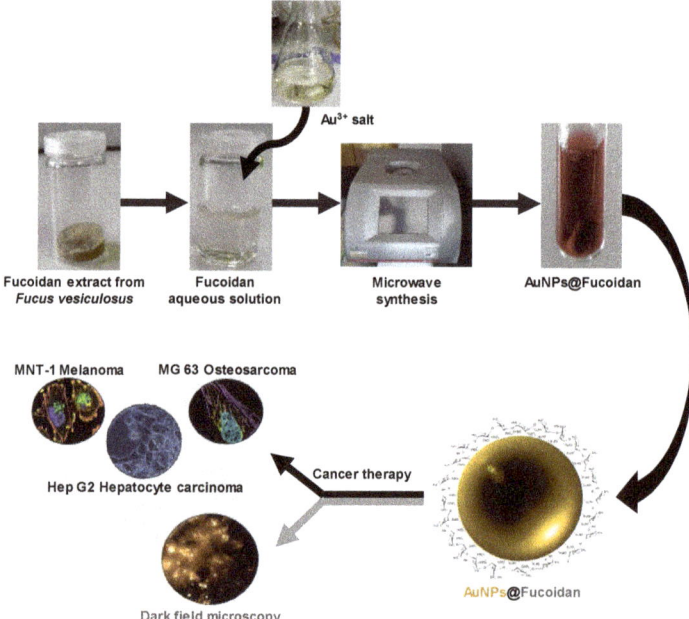

Figure 1. Schematic representation of the microwave irradiated synthesis of fucoidan-AuNPs (molecular structure of fucoidan) with antitumoral activity.

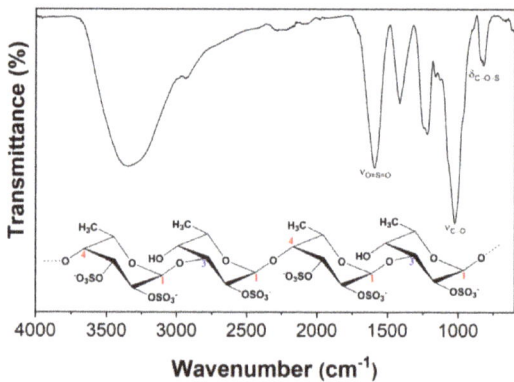

Figure 2. Fourier transform infrared (FTIR) spectrum of the fucoidan rich fraction extracted from *F. vesiculosus* (vibrational modes: ν = stretching, δ = bending).

3.1. Structural and Morphological Characterization of the Fucoidan-AuNPs

The MW-assisted synthesis of AuNPs, using the prepared fucoidan-rich fraction, as the reducing and capping agent, was achieved in only 1 min. Three different concentrations of fucoidan were tested, namely 0.05%, 0.1%, and 0.5% (w/v), aiming to produce AuNPs with an appropriated size and excellent colloidal stability. The color change of the Au solutions from yellowish to ruby red, for fucoidan concentrations of 0.1% and 0.5%, and to purplish for the fucoidan solution with 0.05%, is an early confirmation of the formation of the AuNPs (Figure 3A). The formation of a purple color in the case of the lowest fucoidan concentration (0.05% w/v) could be associated with the formation of larger and close particles due to the lower amount of fucoidan present at the surface of the AuNPs [21]. Additionally, the UV-vis spectra of these fucoidan-AuNPs colloids showed the typical surface plasmon resonance at around 520 nm for fucoidan concentrations of 0.1% and 0.5% (w/v). The position of these bands is indicative of the formation of small spherical NPs. The displacement of this band for higher wavelength values (around 550 nm) for the lowest fucoidan concentration (0.05%) is also in line with the formation of larger particles. Similar results were reported by Jang et al. [21] following a conventional solvothermal method but using considerably higher fucoidan concentrations (from 0.5% to 2.5% w/v).

FTIR analysis of the obtained fucoidan-AuNPs (Figure 3B) clearly confirmed the capping role of fucoidan, because of the occurrence of the typical absorption bands of fucoidan [31], as well as a correlation between the content of fucoidan in the surface of the NPs and its amount used in the synthesis.

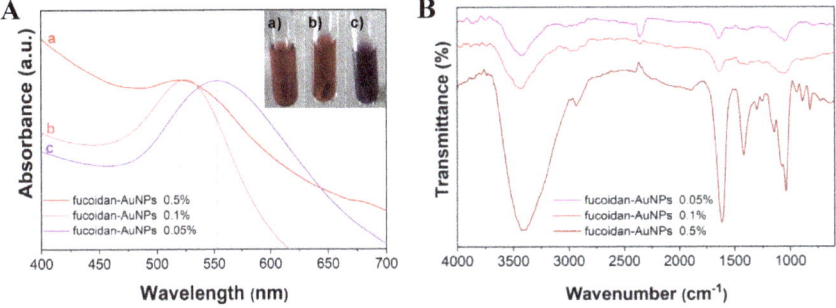

Figure 3. (**A**) UV-Vis and (**B**) FTIR spectra of AuNPs colloids obtained with different concentrations of fucoidan: a) 0.5%, b) 0.1%, and c) 0.05% w/v of fucoidan-rich fraction.

The STEM micrographs provided clear information about the shape and size of the fucoidan-AuNPs (Figure 4). All the obtained fucoidan-AuNPs were monodispersed and spherical, with average sizes of 5.8 ± 0.9 nm, 10.4 ± 1.4 nm, and 13.4 ± 3.0 nm for initial concentrations of fucoidan of 0.5%, 0.1%, and 0.05% (w/v), respectively. STEM images corroborate the increase of the NPs diameter with the decrease of the fucoidan concentration. It is also perceptible that for the fucoidan concentration of 0.05%, AuNPs are closer to each other, but still individualized, because of lower colloidal stability associated with an inferior content of fucoidan on the surface of the NPs. These results are actually remarkable because in conventional solvothermal synthesis, for concentrations equal and lower than 0.5% (w/v) of fucoidan, the AuNPs formed are unstable and aggregate during synthesis leading to large anisotropic particles [21]. These results are a good indication that the MW technology allows the rapid formation of stable spherical AuNPs with controllable size and using lower concentrations of fucoidan when compared with conventional methodologies. Fucoidan-AuNPs sample with an average diameter of 10.4 ± 1.4 nm was selected for the colloidal stability studies and biological evaluations because of the monodispersity and no visual aggregation of the colloidal suspension.

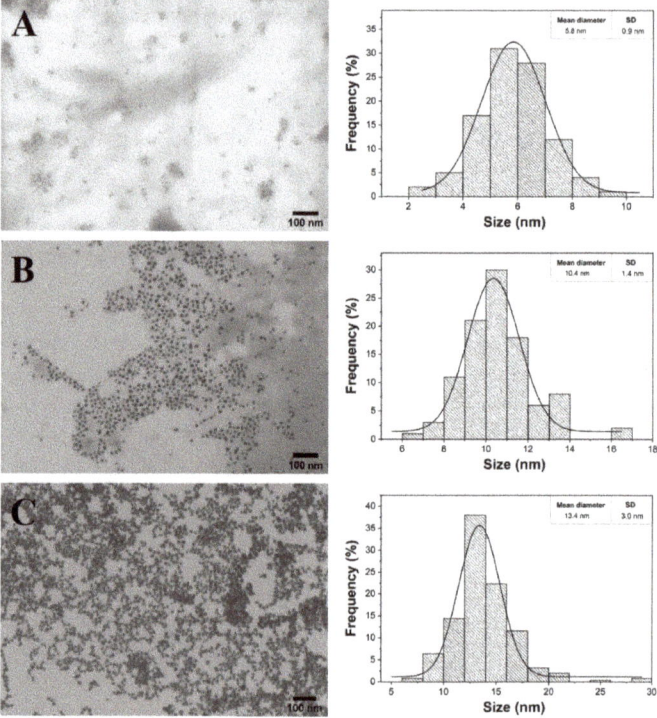

Figure 4. Scanning transmission electron microscopy (STEM) micrographs with the respective histograms of the size distribution of AuNPs colloids obtained with different concentrations of fucoidan: (**A**) 0.5%, (**B**) 0.1%, and (**C**) 0.05% w/v of the fucoidan-rich fraction.

3.2. Colloidal Stability of the Fucoidan-AuNPs

The colloidal stability of the fucoidan-AuNPs (10.4 ± 1.4 nm average diameter) was investigated under different conditions, namely by using acid (pH 2.1) and basic (pH 12) solutions, PBS (pH 7.4), DMEM (culture medium), and ultra-purified water, at room temperature. The stability profile over 48 h was inspected by UV-Vis and STEM analysis. The observation of the fucoidan-AuNPs colloid over time (Figure S1), and for the different conditions allowed to conclude that, in general, these NPs are considerably stable because no color changes were perceived. Based on the relative absorbance maximum obtained in UV-Vis analysis (Figure 5A), it is evident that these fucoidan-AuNPs are highly stable (more than 90% of maximum absorption) in ultra-pure water, DMEM, and alkaline solution (pH 12), with no significant variations over time. Similar results were previously reported for fucoidan-AuNPs with 15–80 nm [21] and fucoidan-Au nanorods [34]. However, in PBS (pH 7.4) and acidic (pH 2.1) solutions, the maximum absorptions slightly decrease over time, reaching about 70% and 80%, respectively. However, the analysis of the colloids by UV-Vis only gives a rough indication of the stability of the NPs because the decrease in the maximum absorption could be associated with different causes. Possible reasons are the change of the refractive index of the surrounding medium and/or the distance between the AuNPs that causes changes in the $\lambda_{máx}$ of absorption (525 nm), as well as in the correspondent absorbance values [35].

To have a deep insight into the effect of the different studied conditions in the stability of the obtained colloids, in particular on their size and morphology, STEM analysis of the AuNPs after 48 h of incubation in the solutions mentioned above was also carried out. STEM micrographs (Figure 5B) confirmed that their morphology and size was not affected, but in acidic conditions, the NPs are much closer. These results indicate that under acidic conditions the fucoidan capping layer is weakened, certainly due to the protonation of sulfate groups and partial detachment from the surface of the AuNPs [36], leading to slightly less stable colloids.

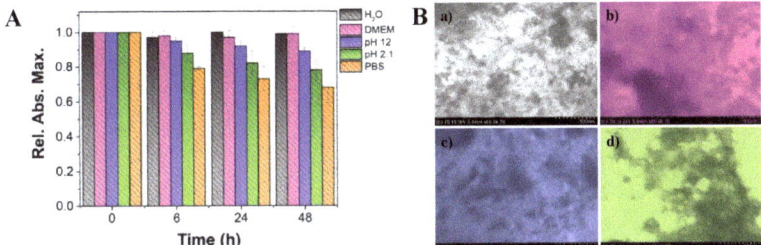

Figure 5. (**A**) Colloidal stability assay of fucoidan-AuNPs 0.1% (w/v) up to 48 h in distinct mediums: acid solution (pH 2.1), basic solution (pH 12), PBS, DMEM (culture medium), and ultra-purified water. (**B**) STEM images of fucoidan-AuNPs after 48 h immersed in the distinct mediums: (**a**) ultra-purified water, (**b**) DMEM, (**c**) NaOH (pH 12), and (**d**) HCl (pH 2.1). The color of micrographs (**b**), (**c**), and (**d**) were changed for visual guidance in order to match the color of the respective medium displayed in (**A**).

3.3. Cytotoxicity Assays of Fucoidan-AuNPs and Cellular Uptake by Flow Cytometry

In this study, the in vitro cytotoxicity of the fucoidan-enriched extract and fucoidan-AuNPs (10.4 ± 1.4 nm) was investigated against MNT-1 (pigmented human melanoma cells), HepG2 (human hepatocyte carcinoma), and MG-63 (human osteosarcoma) cell lines for 24, 48, and 72 h at concentrations ranging from 0 to 5 mg mL^{-1} (Figure 6). The fucoidan extract obtained in this study is not cytotoxic against the three cell lines tested, in the concentration range of 0.25–2.5 mg mL^{-1}, with cell viabilities higher than 90% in most cases. However, for a concentration of 5 mg mL^{-1}, a significant reduction of cell viability (up to 60%) was observed, particularly for 48 and 72 h of exposure. The antitumoral activity of fucoidan (and fucoidan-enriched extracts) is well documented, as well as its dependence on the source of fucoidan [15]. For instance, Manivasagan et al. [19] reported that fucoidan from *F. vesiculosus* inhibits the proliferation of human breast cancer cells with an inhibitory concentration of 35 μg mL^{-1} and Jang et al. [21] described cell viabilities of around 80% for cancer cells (HSC3) treated with 100 mg mL^{-1} of a commercial fucoidan.

The cytotoxic effect of the fucoidan-AuNPs was investigated for concentrations in the range of 2.5–26 μg mL^{-1}. In general, the NPs showed a dose-depend decrease in cell viability but with noticeable differences for distinct cell lines and exposure times. For the HepG2 cell line, it was observed a reduction of cell viability with the concentration of fucoidan-AuNPs reaching about 60% cell viability for 26 μg mL^{-1} of NPs. In this case, no time dependency effect was perceived for the three exposure times investigated. For the other cell lines, it was also observed a decrease in cell viability with the concentration of NPs, but with most pronounced reductions for 48 and 72 h of exposure. For example, for a concentration of 26 μg mL^{-1} of NPs and 72 h of exposure, cell viabilities of 32 and 10% were observed for MNT-1 and MG-63 cell lines, respectively. These results demonstrate the more significant antitumoral effect of fucoidan when combined with the AuNPs because considerably higher cell viability reduction was obtained for much lower concentrations of fucoidan-AuNPs when compared with the fucoidan-enriched extract (around 60% cell viability for 5 mg mL^{-1} of fucoidan). This behavior could be associated with the small size and high surface area of the AuNPs that result in extraordinary surface concentrations of fucoidan and higher interaction with the cells. Jang et al. [21] also reported a higher cell viability reduction for fucoidan-AuNPs comparatively with fucoidan when using the same concentration of 100 mg mL^{-1}.

The cellular uptake of fucoidan-AuNPs was only tested for the MG-63 cells, given the higher antitumoral activity of these nanosystems toward this cell line. According to Figure 7, both concentrations (5 and 12 µg mL^{-1}) of fucoidan-AuNPs induce an increase in side scatter (SS) intensity without change of forward scatter (FS) intensity of MG-63 cells, which means that particles are internalized by the cells.

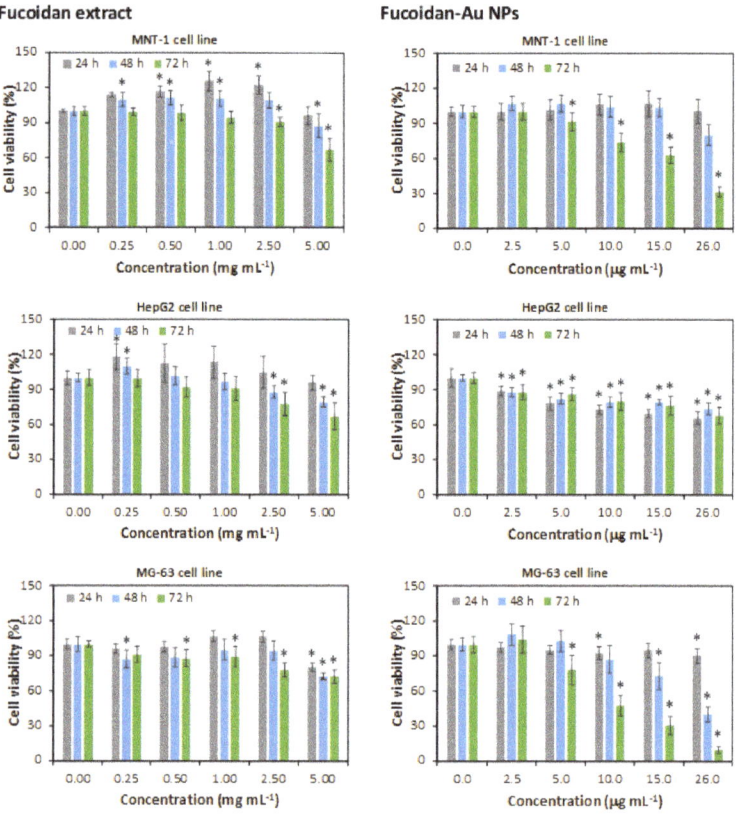

Figure 6. Viability measured by 3-(4,5-dimethylthiazol-2-yl)-2,5-diphenyltetrazolium bromide (MTT) assay after 24, 48, and 72 h exposure to fucoidan extract (left) and fucoidan-AuNPs (right). Values are the mean of nine replicates, and the error bars represent the standard deviation; the asterisk (*) denotes statistically significant differences to the control ($p < 0.05$).

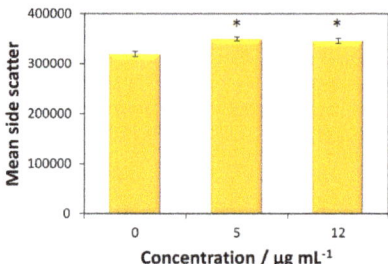

Figure 7. Uptake of fucoidan-AuNPs by MG-63 cells. The uptake was assessed by the side scattered light through flow cytometry after 24 h exposure to 5 and 12 µg mL^{-1} of fucoidan-AuNPs. Values are the mean of three replicates, and the error bars represent the standard deviation; the asterisk (*) denotes statistically significant differences to the control ($p < 0.05$).

3.4. Dark Field Imaging of MG-63 Cells Incubated with Fucoidan-AuNPs

Dark field microscopy of the MG-63 cells and those incubated with fucoidan-AuNPs are shown in Figure 8A,B, respectively. The darker regions in the dark field images are assigned to the cell's nucleus with a diameter of around 20 µm. The cells' nucleus identification is unequivocally confirmed, taking advantage of the fact that the cells were marked with a fluorescent stain (DAPI) that binds specifically to the regions of the nucleus. Thus, under UV irradiation, blue areas are discerned (Figure 8C,D), assigned to the emission spectra around 460 nm of the DAPI (Figure S2). The overlap between the dark field images and those acquired under UV the dark areas overlap that revealing blue emission (Figure 8E). This shows that dark field imaging under white light can be used to identify the nucleus of the cells, without the need to use a fluorescent stain.

Also, the dark field images of the MG-63 cells incubated with fucoidan-AuNPs also shows bright spots with diameter values between 1 and 10 µm. The light scattering from those regions was analyzed by hyperspectral microscopy. Figure 8G,H compare the hyperspectral images of the MG-63 cells and of those incubated with fucoidan-AuNPs, revealing that the bright spots in the images of the MG-63 cells incubated with fucoidan-AuNPs are characterized by a broad spectrum. In fact, it displays a low-relative intensity band in the same region as that found for the absorption of the Au-particles (Figure 3) and a more intense one in the red spectral region, that results from the light scattered by the fucoidan-AuNPs indicating the presence of fucoidan-AuNPs aggregates (Figure 8F, Figures S3 and S4). The larger dimension of those aggregates, when compared to STEM data (Figure 4), is due to the spatial resolution of the optical image, and we also note that the contribution of the guidance of the scattered photons from the particles for the larger bright spots cannot be excluded [37]. We note that some of those bright spots (marked with arrows in Figure 8H) are localized in the same coordinates of the plane in which was possible to detect the cells, suggesting the incorporation of the fucoidan-AuNPs in the MG-63 cells, that it is in line with the flow cytometry results.

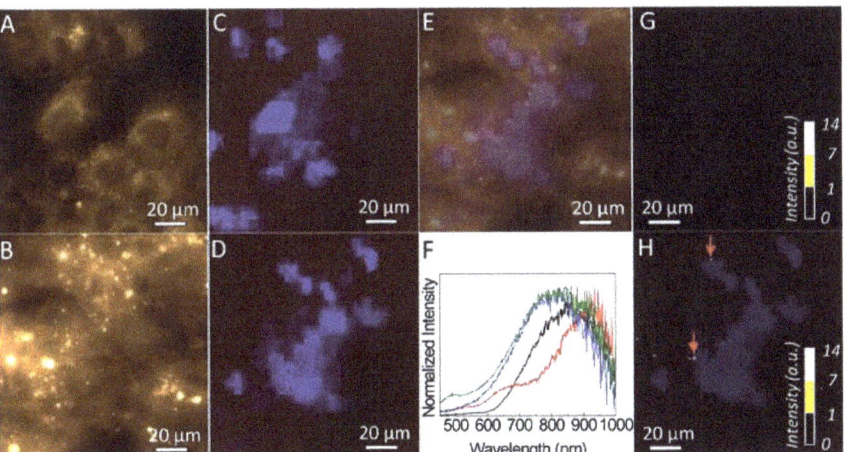

Figure 8. Optical images in dark field transmission mode under white light, of MG-63 cells, incubated (**A**) without and (**B**) with fucoidan-AuNPs. Optical images, in brightfield field reflectance mode under UV irradiation, of MG-63 cells incubated (**C**) without and (**D**) with fucoidan-AuNPs. (**E**) Show the overlay of (**B**) and (**D**). Spectra measured in several single pixels of the bright spots shown in the hyperspectral image shown in the (**F**). (**G**) and (**H**) show the hyperspectral images measured for the same sample and illumination conditions of (A) and (B), respectively. The color scale is based on the intensity of the spectra of each pixel at 750 nm. In (H), the hyperspectral image is superimposed with (D).

4. Conclusions

The one-minute microwave-assisted synthesis of fucoidan-coated gold nanoparticles (AuNPs) with controllable, high stability, and antitumoral activity is the first and foremost contribution of the present study, where fucoidan-AuNPs were synthesized by using a fucoidan-enriched fraction extracted from *F. vesiculosus*, as simultaneous reducing and capping agent. The resulting monodispersed and spherical fucoidan-AuNPs present very small diameters, namely 5.8 ± 0.9 nm, 10.4 ± 1.4 nm, and 13.4 ± 3.0 nm that depend on the initial concentrations of fucoidan: 0.5%, 0.1%, and 0.05% (w/v), respectively, together with excellent colloidal stability in acidic and basic solutions, ultra-purified water and culture media. The second innovative input of this study lies in the antitumoral activity of the fucoidan-AuNPs against three human tumor cell lines, namely MNT-1 (pigmented human melanoma), HepG2 (human hepatocyte carcinoma), and MG-63 (human osteosarcoma) cells. Moreover, the use of flow cytometry in combination with dark field imaging confirmed the cellular uptake of fucoidan-AuNPs by the MG-63 cell line, which demonstrates the antitumoral activity of these nanomaterials, and thus their potential for cancer therapy.

Supplementary Materials: The following are available online at http://www.mdpi.com/1996-1944/13/5/1076/s1. Figure S1: Digital photographs of fucoidan-AuNPs colloid over time in the different media. Figure S2: Hyperspectral images of MG-63 cells incubated (**A**) without and (**B**) with fucoidan-AuNPs, measured for the same sample and illumination conditions of Figure 8C,D of the manuscript. The color scale is based on the intensity of the spectra of each pixel at 460 nm. (**C**) Spectra measured in collection areas of 20 × 20 pixels of (A) (red line) and (B) (blue line). Figure S3: Spectra measured in several single pixels of the hyperspectral image measured for the same sample and illumination conditions of the Figure 8A of the manuscript, before (**A**) and after (**C**) correction. (**B**) Spectrum measured in an area of 20 × 20 pixels, of the Figure 8A of the manuscript used to perform the correction. The inset in (A) shows the normalized spectra for better comparison. Figure S4: Spectra measured in several single pixels of the hyperspectral image shown in Figure 8B of the manuscript, (**A**) before and (**C**) after correction using the spectrum shown in Figure S3B. The normalized spectra before correction are shown in (**B**) for better comparison.

Author Contributions: Conceptualization, R.J.B.P., S.A.O.S., and C.S.R.F.; investigation, R.J.B.P., D.B., C.V., A.M.P.B., A.C.M., and F.C.; resources, R.A.S.F., H.O., M.H.A., S.A.O.S., and C.S.R.F.; writing—original draft preparation, C.S.R.F.; writing—review and editing, R.J.B.P., D.B., C.V., A.M.P.B., R.A.S.F., A.C.M., F.C., H.O., M.H.A., S.A.O.S., and C.S.R.F.; supervision, C.S.R.F.; funding acquisition, R.A.S.F., H.O., S.A.O.S., and C.S.R.F. All authors have read and agreed to the published version of the manuscript.

Funding: This work was developed within the scope of the project CICECO-Aveiro Institute of Materials, UIDB/50011/2020 & UIDP/50011/2020, and CESAM, UID/AMB/50017/2019, financed by national funds through the FCT/MEC and when appropriate co-financed by FEDER under the PT2020 Partnership Agreement. The research contract of R.J.B. Pinto is funded by national funds (OE), through FCT–Fundação para a Ciência e a Tecnologia, I.P., in the scope of the framework contract foreseen in the numbers 4, 5 and 6 of the article 23, of the Decree-Law 57/2016, of August 29, changed by Law 57/2017, of July 19. The research contract of S.A.O. Santos is funded by the project AgroForWealth (CENTRO-01-0145-FEDER-000001). FCT is also acknowledged for the research contracts under Scientific Employment Stimulus to C. Vilela (CEECIND/00263/2018), H. Oliveira (CEECIND/04050/2017) and C.S.R. Freire (CEECIND/00464/2017). A.M.P. Botas acknowledges SusPhotoSolutions (CENTRO-01-0145-FEDER-000005).

Conflicts of Interest: The authors declare no conflict of interest.

References

1. World Health Organization. Available online: https://www.who.int/cancer/en/ (accessed on 6 June 2019).
2. Roy Chowdhury, M.; Schumann, C.; Bhakta-Guha, D.; Guha, G. Cancer nanotheranostics: Strategies, promises and impediments. *Biomed. Pharmacother.* **2016**, *84*, 291–304. [CrossRef] [PubMed]
3. Mitchell, M.J.; Jain, R.K.; Langer, R. Engineering and physical sciences in oncology: Challenges and opportunities. *Nat. Rev. Cancer* **2017**, *17*, 659–675. [CrossRef] [PubMed]
4. Chen, H.; Zhang, W.; Zhu, G.; Xie, J.; Chen, X. Rethinking cancer nanotheranostics. *Nat. Rev. Mater.* **2017**, *2*, 17024. [CrossRef] [PubMed]
5. Aftab, S.; Shah, A.; Nadhman, A.; Kurbanoglu, S.; Aysıl Ozkan, S.; Dionysiou, D.D.; Shukla, S.S.; Aminabhavi, T.M. Nanomedicine: An effective tool in cancer therapy. *Int. J. Pharm.* **2018**, *540*, 132–149. [CrossRef]

6. Silva, C.; Pinho, J.; Lopes, J.; Almeida, A.; Gaspar, M.; Reis, C. Current Trends in Cancer Nanotheranostics: Metallic, Polymeric, and Lipid-Based Systems. *Pharmaceutics* **2019**, *11*, 22. [CrossRef]
7. Arvizo, R.R.; Bhattacharyya, S.; Kudgus, R.A.; Giri, K.; Bhattacharya, R.; Mukherjee, P. Intrinsic therapeutic applications of noble metal nanoparticles: Past, present and future. *Chem. Soc. Rev.* **2012**, *41*, 2943–2970. [CrossRef]
8. Menon, J.U.; Jadeja, P.; Tambe, P.; Vu, K.; Yuan, B.; Nguyen, K.T. Nanomaterials for Photo-Based Diagnostic and Therapeutic Applications. *Theranostics* **2013**, *3*, 152–166. [CrossRef]
9. Li, W.; Chen, X. Gold nanoparticles for photoacoustic imaging. *Nanomedicine* **2015**, *10*, 299–320. [CrossRef]
10. Weber, J.; Beard, P.C.; Bohndiek, S.E. Contrast agents for molecular photoacoustic imaging. *Nat. Methods* **2016**, *13*, 639–650. [CrossRef] [PubMed]
11. Abadeer, N.S.; Murphy, C.J. Recent Progress in Cancer Thermal Therapy Using Gold Nanoparticles. *J. Phys. Chem. C* **2016**, *120*, 4691–4716. [CrossRef]
12. Her, S.; Jaffray, D.A.; Allen, C. Gold nanoparticles for applications in cancer radiotherapy: Mechanisms and recent advancements. *Adv. Drug Deliv. Rev.* **2017**, *109*, 84–101. [CrossRef] [PubMed]
13. Nam, J.; Won, N.; Bang, J.; Jin, H.; Park, J.; Jung, S.; Jung, S.; Park, Y.; Kim, S. Surface engineering of inorganic nanoparticles for imaging and therapy. *Adv. Drug Deliv. Rev.* **2013**, *65*, 622–648. [CrossRef] [PubMed]
14. Yadav, P.; Singh, S.P.; Rengan, A.K.; Shanavas, A.; Srivastava, R. Gold laced bio-macromolecules for theranostic application. *Int. J. Biol. Macromol.* **2018**, *110*, 39–53. [CrossRef] [PubMed]
15. Wu, L.; Sun, J.; Su, X.; Yu, Q.; Yu, Q.; Zhang, P. A review about the development of fucoidan in antitumor activity: Progress and challenges. *Carbohydr. Polym.* **2016**, *154*, 96–111. [CrossRef]
16. Fitton, J.; Stringer, D.; Karpiniec, S. Therapies from Fucoidan: An Update. *Mar. Drugs* **2015**, *13*, 5920–5946. [CrossRef]
17. Manivasagan, P.; Oh, J. Marine polysaccharide-based nanomaterials as a novel source of nanobiotechnological applications. *Int. J. Biol. Macromol.* **2016**, *82*, 315–327. [CrossRef]
18. Tengdelius, M.; Gurav, D.; Konradsson, P.; Påhlsson, P.; Griffith, M.; Oommen, O.P. Synthesis and anticancer properties of fucoidan-mimetic glycopolymer coated gold nanoparticles. *Chem. Commun.* **2015**, *51*, 8532–8535. [CrossRef]
19. Manivasagan, P.; Bharathiraja, S.; Bui, N.Q.; Jang, B.; Oh, Y.O.; Lim, I.G.; Oh, J. Doxorubicin-loaded fucoidan capped gold nanoparticles for drug delivery and photoacoustic imaging. *Int. J. Biol. Macromol.* **2016**, *91*, 578–588. [CrossRef]
20. Kim, H.; Nguyen, V.P.; Manivasagan, P.; Jung, M.J.; Kim, S.W.; Oh, J.; Kang, H.W. Doxorubicin-fucoidan-gold nanoparticles composite for dual-chemo-photothermal treatment on eye tumors. *Oncotarget* **2017**, *8*, 113719–113733. [CrossRef]
21. Jang, H.; Kang, K.; El-Sayed, M.A. Facile size-controlled synthesis of fucoidan-coated gold nanoparticles and cooperative anticancer effect with doxorubicin. *J. Mater. Chem. B* **2017**, *5*, 6147–6153. [CrossRef]
22. Arshi, N.; Ahmed, F.; Kumar, S.; Anwar, M.; Lu, J.; Heun Koo, B.; Gyu Lee, C. Microwave assisted synthesis of gold nanoparticles and their antibacterial activity against Escherichia coli (*E. coli*). *Curr. Appl. Phys.* **2011**, *11*, S360–S363. [CrossRef]
23. Bayazit, M.K.; Yue, J.; Cao, E.; Gavriilidis, A.; Tang, J. Controllable Synthesis of Gold Nanoparticles in Aqueous Solution by Microwave Assisted Flow Chemistry. *ACS Sustain. Chem. Eng.* **2016**, *4*, 6435–6442. [CrossRef]
24. Rodriguez-Jasso, R.M.; Mussatto, S.; Pastrana, L.; Aguilar, C.; Teixeira, J. Chemical composition and antioxidant activity of sulphated polysaccharides extracted from Fucus vesiculosus using different hydrothermal processes. *Chem. Pap.* **2014**, *68*, 203–209. [CrossRef]
25. Rioux, L.-E.; Turgeon, S.L.; Beaulieu, M. Rheological characterisation of polysaccharides extracted from brown seaweeds. *J. Sci. Food Agric.* **2007**, *87*, 1630–1638. [CrossRef]
26. Roger, O.; Kervarec, N.; Ratiskol, J.; Colliec-Jouault, S.; Chevolot, L. Structural studies of the main exopolysaccharide produced by the deep-sea bacterium Alteromonas infernus. *Carbohydr. Res.* **2004**, *339*, 2371–2380. [CrossRef] [PubMed]
27. DuBois, M.; Gilles, K.A.; Hamilton, J.K.; Rebers, P.A.; Smith, F. Colorimetric Method for Determination of Sugars and Related Substances. *Anal. Chem.* **1956**, *28*, 350–356. [CrossRef]
28. Twentyman, P.; Luscombe, M. A study of some variables in a tetrazolium dye (MTT) based assay for cell growth and chemosensitivity. *Br. J. Cancer* **1987**, *56*, 279–285. [CrossRef]

29. Suzuki, H.; Toyooka, T.; Ibuki, Y. Simple and Easy Method to Evaluate Uptake Potential of Nanoparticles in Mammalian Cells Using a Flow Cytometric Light Scatter Analysis. *Environ. Sci. Technol.* **2007**, *41*, 3018–3024. [CrossRef]
30. Bastos, V.; Ferreira-de-Oliveira, J.M.P.; Carrola, J.; Daniel-da-Silva, A.L.; Duarte, I.F.; Santos, C.; Oliveira, H. Coating independent cytotoxicity of citrate- and PEG-coated silver nanoparticles on a human hepatoma cell line. *J. Environ. Sci.* **2017**, *51*, 191–201. [CrossRef]
31. Rodriguez-Jasso, R.M.; Mussatto, S.I.; Pastrana, L.; Aguilar, C.N.; Teixeira, J.A. Microwave-assisted extraction of sulfated polysaccharides (fucoidan) from brown seaweed. *Carbohydr. Polym.* **2011**, *86*, 1137–1144. [CrossRef]
32. Ale, M.T.; Mikkelsen, J.D.; Meyer, A.S. Important determinants for fucoidan bioactivity: A critical review of structure-function relations and extraction methods for fucose-containing sulfated polysaccharides from brown seaweeds. *Mar. Drugs* **2011**, *9*, 2106–2130. [CrossRef] [PubMed]
33. Pielesz, A.; Biniaś, W. Cellulose acetate membrane electrophoresis and FTIR spectroscopy as methods of identifying a fucoidan in Fucus vesiculosus Linnaeus. *Carbohydr. Res.* **2010**, *345*, 2676–2682. [CrossRef] [PubMed]
34. Manivasagan, P.; Bharathiraja, S.; Santha Moorthy, M.; Oh, Y.O.; Song, K.; Seo, H.; Oh, J. Anti-EGFR Antibody Conjugation of Fucoidan-Coated Gold Nanorods as Novel Photothermal Ablation Agents for Cancer Therapy. *ACS Appl. Mater. Interfaces* **2017**, *9*, 14633–14646. [CrossRef] [PubMed]
35. Liz-Marzán, L.M. Nanometals. *Mater. Today* **2004**, *7*, 26–31. [CrossRef]
36. Santha Moorthy, M.; Subramanian, B.; Panchanathan, M.; Mondal, S.; Kim, H.; Lee, K.D.; Oh, J. Fucoidan-coated core–shell magnetic mesoporous silica nanoparticles for chemotherapy and magnetic hyperthermia-based thermal therapy applications. *New J. Chem.* **2017**, *41*, 15334–15346. [CrossRef]
37. Gonell, F.; Botas, A.M.P.; Brites, C.D.S.; Amorós, P.; Carlos, L.D.; Julián-López, B.; Ferreira, R.A.S. Aggregation-induced heterogeneities in the emission of upconverting nanoparticles at the submicron scale unfolded by hyperspectral microscopy. *Nanoscale Adv.* **2019**, *1*, 2537–2545. [CrossRef]

 © 2020 by the authors. Licensee MDPI, Basel, Switzerland. This article is an open access article distributed under the terms and conditions of the Creative Commons Attribution (CC BY) license (http://creativecommons.org/licenses/by/4.0/).

Article

Noble Metal Composite Porous Silk Fibroin Aerogel Fibers

Alexander N. Mitropoulos [1,2,*], F. John Burpo [1,*], Chi K. Nguyen [1], Enoch A. Nagelli [1], Madeline Y. Ryu [1], Jenny Wang [1], R. Kenneth Sims [3], Kamil Woronowicz [1] and J. Kenneth Wickiser [1]

[1] Department of Chemistry and Life Science, United States Military Academy, West Point, NY 10996, USA; chi.nguyen@westpoint.edu (C.K.N.); enoch.nagelli@westpoint.edu (E.A.N.); madeline.ryu@westpoint.edu (M.Y.R.); jenny.wang@westpoint.edu (J.W.); kamil.woronowicz@westpoint.edu (K.W.); ken.wickiser@westpoint.edu (J.K.W.)
[2] Department of Mathematical Sciences, United States Military Academy, West Point, NY 10996, USA
[3] Department of Civil and Mechanical Engineering, United States Military Academy, West Point, NY 10996, USA; robert.sims@westpoint.edu
* Correspondence: alexander.mitropoulos@gmail.com (A.N.M.); john.burpo@westpoint.edu (F.J.B.); Tel.: +1-845-938-3900 (F.J.B)

Received: 11 February 2019; Accepted: 11 March 2019; Published: 18 March 2019

Abstract: Nobel metal composite aerogel fibers made from flexible and porous biopolymers offer a wide range of applications, such as in catalysis and sensing, by functionalizing the nanostructure. However, producing these composite aerogels in a defined shape is challenging for many protein-based biopolymers, especially ones that are not fibrous proteins. Here, we present the synthesis of silk fibroin composite aerogel fibers up to 2 cm in length and a diameter of ~300 µm decorated with noble metal nanoparticles. Lyophilized silk fibroin dissolved in hexafluoro-2-propanol (HFIP) was cast in silicon tubes and physically crosslinked with ethanol to produce porous silk gels. Composite silk aerogel fibers with noble metals were created by equilibrating the gels in noble metal salt solutions reduced with sodium borohydride, followed by supercritical drying. These porous aerogel fibers provide a platform for incorporating noble metals into silk fibroin materials, while also providing a new method to produce porous silk fibers. Noble metal silk aerogel fibers can be used for biological sensing and energy storage applications.

Keywords: biopolymer; silk fibroin; aerogel; fiber; nanomaterials; nanoparticles; noble metals; gold; platinum; palladium

---

1. Introduction

Biopolymers provide unique applications in advanced technology where degradation combined with natural materials are required. In nature, biopolymers take several forms, such as films, fibers, gels, and sponges, which are optimized for their required applications [1,2]. However, producing the equivalent forms with the desired qualities in regenerated biopolymers has been challenging, especially making fibers with controlled diameters and porosity [2]. Nevertheless, working with regenerated biopolymer solutions can enhance properties such as the tensile strength and porosity [3]. Furthermore, regenerated biopolymer solutions can be used as a structural network that can be combined with other materials to synthesize composites not found in nature [4,5]. Starting with biopolymers dissolved in polar solvents, such as water or alcohols, can be useful in providing varied biopolymer conformational folding to enhance the desired properties.

Silk fibroin, purified from the *Bombyx mori* silk caterpillar, is a well-established protein that is processable into fibers, [6,7] films, [8–11] foams, particles, [12] hydrogels, [13,14] and, recently, aerogels after supercritical drying with CO_2 [15–17]. The production of silk fibroin-based materials requires a detailed understanding of the solvent-mediated dielectric environment to induce the hierarchical self-assembly responsible for the mechanical properties resulting from protein primary structure and molecular assembly [18]. Particularly, in silk fibroin, there are three major folding secondary structure motifs, including random coils, alpha helices, or beta-sheets, which control the silk's strength [19]. Specifically, silk fibroin molecular folding can be controlled by the solvent, which forces its primary structure to arrange into the above mentioned secondary structures [18,20].

Alcohols as solvents are useful to stabilize specific secondary structures of silk fibroin in aqueous environments while denaturing the native tertiary conformation [21]. Because the properties of silk fibroin depend strongly on the preparation conditions (fluid environment), the choice of solvent affects the overall quality of the bulk material and the formed nanostructure [3]. 1,1,1,3,3,3-Hexafluoro-propan-2-ol (HFIP) is one of the most applicable solvents for the stabilization of silk's secondary structures, particularly the alpha-helical conformation [21–26]. The high polarity of HFIP as an alcohol allows it to stabilize the silk helical state by decreasing the polarity of the protein chains. This results in local hydrogen bonds that stabilize amphiphilic helical conformations, producing a silk alcogel [21]. The protein concentration is a significant factor in unfolded conformations as the silk secondary structure switches between random coil and alpha-helix depending on the different molecular interactions that can occur [21].

Silk fibroin exposed to HFIP or other alcohols has been used to make gel materials, which can be used for fracture fixation systems or artificial fibers after convective drying [18,26,27]. Dissolving silk in HFIP to cast different forms, particularly gel materials, before convective drying confers the mechanical properties of the folded protein structure and maintains the assembled bulk solid structure in the alcogel formation [27]. Additionally, controlling the nanostructure of the silk fibroin, nanofibrils has been achieved with supercritical CO_2 drying ($SCCO_2$) [17]. Supercritical drying ensures the porosity of the material and maintains the high surface area and low density [28–30]. Using this drying method results in molecular conformational changes of the HFIP–silk fiber, which produce stronger, high surface area materials that have potential for novel biomedical applications.

Furthermore, enhancing the properties of biopolymer gels, such as metallization for catalysis and sensing, can be achieved by equilibrating with gold, palladium, and platinum noble metal complexes, which, after electrochemical reduction, result in nanoparticle growth on the biopolymer nanofibrils [4,5]. Using silk fibroin as the material of choice to add conductive noble metal nanoparticles would enhance the versatility of this mechanically robust and biocompatible material.

Here, we demonstrate the preparation of a composite material consisting of noble metal nanoparticles attached to HFIP-treated porous silk aerogel fibers forming a composite material. Maintaining a constant concentration but changing the type of noble metal ion species determines the extent of the nanoparticle growth on the silk nanofibril surface along with the resulting percentage of metal content in the final aerogel fiber. This allows for material variation based on nobility compared with concentration. The controllable bulk geometries improve this platform to allow it to be used for biofibers. Lastly, utilizing different noble metals can extend the applicability of these biopolymer nanofibrils for other applications, such as catalysis, energy storage, and sensing.

2. Materials and Methods

2.1. Silk Fibroin Fiber Aerogel Synthesis

Silk fibroin solution was prepared as previously described [31]. *B. mori* silkworm cocoons were boiled for 30 minutes in a solution of 0.02 M Na_2CO_3 to remove the sericin glycoprotein. The extracted fibroin was rinsed in deionized water and dried at ambient conditions for 12 h. The dried fibroin was dissolved in 9.3 M LiBr solution at 60 °C for 3 h. The solution was dialyzed against deionized water using a dialysis cassette (Slide-a-Lyzer, Pierce, molecular weight cut-off (MWCO) 3.5 kDa) at room temperature for 2 days until the solution reached a concentration of approximately 60 mg/mL. The obtained solution was purified by centrifugation (20 min at 11,000× g) to remove large aggregates.

For the HFIP aerogels, reconstituted silk fibroin was frozen and lyophilized, resulting in a dried material. The dried silk was stored in ambient conditions to prevent any rehydration of the lyophilized silk fibroin. The lyophilized silk was resolubilized in 1,1,1,3,3,3-Hexafluoro-propan-2-ol (HFIP) (Matrix Scientific, Columbia, SC, USA) to generate a 40 mg/mL solution. The concentration is critical as solutions with lower concentrations produce aggregated gels. The solubilized HFIP silk solution was stored at ambient temperatures until used in the gel-forming process.

The silk aerogels were prepared using silicon tubing with an inner diameter of 1.5 mm (McMaster-Carr, Robbinsville, NJ, USA) and filled with HFIP/silk solution. After the HFIP/silk solution was put into the silicon tubing, the HFIP-silk was submerged in 200 proof ethanol (Fisher Scientific, Waltham, MA, USA). Ethanol diffusion into the HFIP-silk proceeded for 24 h to induce physical crosslinking and form a free-standing fiber gel. Additional ethanol rinses were performed to displace the HFIP after the silk gel fiber was removed from the silicon tubing.

After physical crosslinking, the silk fiber gels were rinsed in deionized water and equilibrated in 100 mM of sodium tetrachloropalladate (II) (Na_2PdCl_4), potassium tetrachloroplatinate (II) (K_2PtCl_4), or gold chloride trihydrate ($HAuCl_4 \cdot 3H_2O$) (Sigma Aldrich, St. Louis, MO, USA) for 24 h.

The fiber samples were reduced in 100 mM sodium borohydride for 24 h for noble metal nanoparticle growth [32]. The silk–metal composite gels were rinsed in deionized water for 24 h to remove excess reducing agent. To maintain the metal-coated nanofibrillar hydrogel network, samples were then dehydrated in a series of ethanol rinses at concentrations of 25, 50, 75, and 100% for 30 min each and then supercritically dried in CO_2 using a Leica EM CPD300 Automated Critical Point Dryer (Buffalo Grove, IL, USA) with a set point of 35 °C and 1200 psi.

To prepare sufficient sample material for X-ray diffraction, thermal gravimetric, and nitrogen gas adsorption analysis, silk–metal composites were prepared in a bulk monolith geometry. A HFIP–silk solution was cast in a 48-well cell culture dish (diameter of 10 mm) and crosslinked with 200 proof ethanol by casting on the top for 24 h. After crosslinking, silk hydrogels were rinsed in deionized water for 48 h to remove any remaining HFIP and ethanol. The hydrogels were equilibrated in 100 mM of sodium tetrachloropalladate (II) (Na_2PdCl_4), potassium tetrachloroplatinate (II) (K_2PtCl_4), or gold chloride trihydrate ($HAuCl_4 \cdot 3H_2O$) as in the case of the silk–metal fiber synthesis above but for 48 h. To ensure reduction throughout the volume of the silk gels, 2 M sodium borohydride ($NaBH_4$) and 2 M dimethylamine borane (DMAB) was used for the palladium-, and platinum- and gold-equilibrated silk gels, respectively. The high reducing agent concentration was used to drive diffusion into the depth of the gel. Electrochemical reduction proceeded for 24 h before rinsing for 48 h in deionized water. An ethanol solvent exchange was performed prior to supercritical drying in CO_2.

2.2. Scanning Electron Microscopy (SEM)

SEM was used to evaluate scaffold morphology. All the micrographs were taken with a TM-3000 Scanning Electron Microscope (Hitachi, Tokyo, Japan) or a FEI Helios 600 scanning electron microscope (ThermoFisher Scientific, Hillsboro, OR, USA). Samples were not coated in gold prior to imaging.

2.3. X-ray Diffractometry (XRD)

XRD measurements were performed using a PANalytical Empyrean (Malvern PANalytical, Almelo, The Netherlands) diffractometer with scans at 45 kV and 40 mA with Cu K_α radiation (1.54060 Å), a 2θ step size of 0.0130°, and 20 s per step for diffraction angles (2θ) performed from 5° to 90°. XRD spectra analysis was performed with High Score Plus software (Malvern PANalytical, Almelo, The Netherlands). Crystallite size (D) was determined with the Debeye–Scherrer formula $D = K\lambda(B\cos\theta)^{-1}$ with the shape factor (K), full width at half maxima (B), radiation wave length (λ), and Bragg angle (θ). A shape factor of K = 0.9 was used. High Score Plus software was used to analyze the XRD spectra (Version 4.6, Malvern PANalytical, Almelo, The Netherlands).

2.4. Thermal Gravimetric Analysis (TGA)

Thermal gravimetric analysis (TGA) was performed on a Thermal Instruments Q-500 (New Castle, DE, USA) in a ramp state with a temperature rate of 10 °C/min from ambient to 1000 °C. Samples were maintained under nitrogen gas flow at a rate of 60 ml/min.

2.5. Fourier Transform Infrared (FTIR) Spectroscopy

FTIR analysis of silk film samples was performed in a PerkinElmer Frontier Optica FIR spectrometer (PerkinElmer, Waltham, MA, USA) in attenuated total reflectance (ATR). Films were measured before and after the galvanic displacement. For each sample, 64 scans were collected with a resolution of 1 cm^{-1}, with a wave number range of 4000–650 cm^{-1}.

2.6. Porosity and Surface Area Analysis

Nitrogen gas adsorption–desorption measurements were performed according to International Union of Pure and Applied Chemistry (IUPAC) standards [33] using a Micromeritics ASAP 2020 Plus (Micromeritics, Norcross, GA, USA) to determine surface area and pore size. All the samples were vacuum degassed at 100 °C for 10 h prior to analysis. Brunauer–Emmett–Teller (BET) analysis [34] was used to determine the specific surface area from gas adsorption. Pore size distributions for each sample were calculated using the Barrett–Joyner–Halenda (BJH) model [35] applied to volumetric desorption isotherms. All the calculations were performed using Micromeritics' ASAP 2020 software (Micromeritics, Norcross, GA, USA).

3. Results and Discussion

3.1. Silk Fibroin Aerogel Synthesis

A fast and robust method was developed to form noble metal composite silk fibroin nanostructured fibers by crosslinking HFIP–silk solution with ethanol. The average molecular weight of the silk was 100 kDa and a concentration of 40 mg/mL. Figure 1 shows the synthesis scheme for the silk fibroin noble metal composite aerogels. In order to physically crosslink the silk, the HFIP–silk solution was injected into a silicon tubing mold with an inner diameter of 1.5 mm and 3 cm long and submerged in a bath of 100% ethanol (Figure 1c–d). Parameters such as silk fibroin concentration, noble metal concentration, and reducing agent concentration were the determining factors in the composite fibers. The type of noble metal dictated the noble metal particle size and allowed for particles less than 10 nm and noble metal deposition onto the silk template. Visible deposition of noble metal nanoparticles was observed by a visible color change compared with a control (Figure 1g–h). The formed fibers after reduction provided flexibility and were able to bend into geometric shapes.

To demonstrate the ability to deposit multiple noble metals, silk hydrogels were prepared as cylindrical gels and equilibrated in 100 mM noble metal ion solutions of palladium (Na_2PdCl_4) and platinum (K_2PtCl_6) (Figure S1). Reduction of the metal ions can occur from multiple reducing agents, such as DMAB or $NaBH_4$, as represented by palladium reduction and platinum reduction in Figure S1c,d, respectively. Furthermore, 100 mM metal ion solutions of gold ($HAuCl_4$) were also reduced in DMAB (Figure S2a–c). The noble metal type and reducing agent changes the morphology of the aerogel fibers as visible in SEM images (Figures S3 and S4). The silk–gold aerogels reduced in DMAB show thinner nanofibrils (30–50 nm) with smaller pore sizes compared with the silk–platinum aerogels reduced in $NaBH_4$, which have larger diameter nanofibrils (70–100 nm) and larger pores.

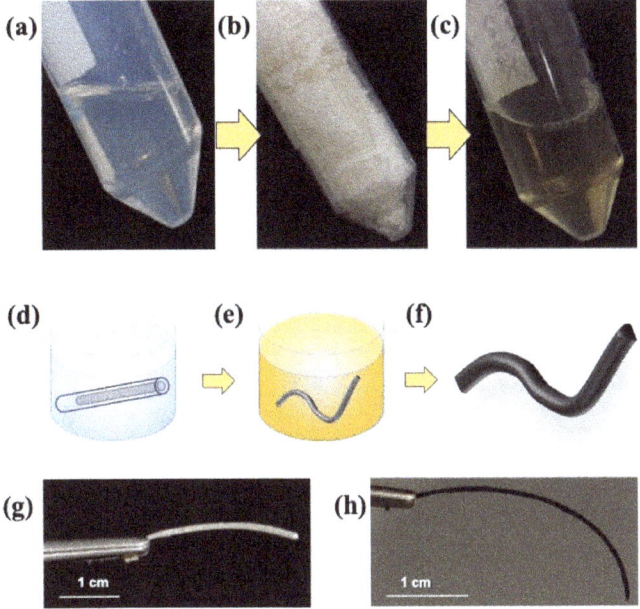

Figure 1. Silk fibroin aerogel fiber synthesis scheme. Different working solutions of silk fibroin starting with (**a**) regenerated silk fibroin solution, (**b**) lyophilized silk fibroin, and (**c**) hexafluoro-2-propanol (HFIP)–silk fibroin. Scheme depiction of the synthesis of noble metal silk fibroin aerogel fibers. (**d**) HFIP–silk fibroin in silicon tubing mold in an ethanol bath to induce physical crosslinking, (**e**) equilibrating in noble metal ionic solution, and (**f**) after reduction and supercritical drying. (**g**) The silk fibroin aerogel fiber without noble metal addition and (**h**) the silk–palladium aerogel fiber (scale bars are 1 cm).

3.2. Aerogel Morphology and Noble Metal Nanoparticles

The structure of the nanofibrils and growth of the noble metal nanoparticles indicates the effects of reduction on the silk fibroin. It has been previously demonstrated that the reduction of these noble metals can be completed using gelatin and cellulose materials with higher reducing agent concentrations [4,5]. Figure 2 shows the SEM images of the silk fibroin aerogel with platinum and palladium after reduction with 100 mM sodium borohydride. Figure 2a shows the fiber formation of the HFIP–silk aerogel fiber. At higher magnification, the presence of nanoparticles covers the silk fiber network for the silk–palladium composite aerogel fibers (Figure 2b–c). This is observed by the brighter regions indicative of nanoparticle growth. The silk–platinum composite aerogel fibers show a more even distribution of noble metal nanoparticle growth onto the silk fibroin (Figure 2d–e).

The morphology of the underlying HFIP–silk fibroin template shows an interconnected network of silk protein spherical particles with diameters of ~500 nm forming a string of pearl-like fibers with high porosity, which is caused by the gelation of the original HFIP–silk (Figure S3). The interconnected network is visible at higher magnification in the anchored silk–gold composite aerogel fibers (Figure S4a–c) when reduced with DMAB. The gold nanoparticles cluster around the larger silk nanofibrils in discrete nanoparticles. Chemical reduction with sodium borohydride of the silk–platinum composite aerogels shows an interconnected silk nanofibril network with nanoparticles dispersed on the surface with a diameter in the range of 5–20 nm (Figure S4d–f).

Figure 2. Scanning electron microscopy images. (**a–c**) The silk–palladium composite aerogel fibers. (**d–e**) The silk–platinum composite aerogel fibers.

The change in fiber morphology is possibly caused by the changes in pH of the noble metal ion solution where solutions of HAuCl$_4$ have lower pH values compared with K$_2$PtCl$_6$ or Na$_2$PdCl$_4$. Due to the lower pH, gold metal ions are more attracted to the silk molecular structure caused by electrostatic interactions, which is visible as a higher density of decorated surface nanoparticles compared with the silk–platinum and silk–palladium aerogels (Figure S3). This has been discussed in other biopolymer-related work [4,5]. Silk fibroin has a repeating pattern of glycine, serine, and alanine forming polymer blocks with shorter block regions of non-repeating sequences [36]. It is these amino acids that allow silk to grow noble metal nanoparticles.

3.3. XRD Characterization

Figure 3a shows the X-ray diffraction (XRD) spectra for the HFIP–silk composite aerogel composites synthesized with palladium and platinum, respectively. XRD spectra for aerogel composites prepared with palladium were indexed to the Joint Committee on Powder Diffraction Standards (JCPDS) reference number 01-087-0637 for palladium and 01-073-0004 for palladium hydride. For aerogel composites prepared with platinum, XRD peaks were indexed to JCPDS reference number 00-004-0802 for platinum. Both palladium and platinum phases are cubic crystal systems with Fm-3 m space groups. The shape evolution of the palladium and platinum nanocrystals of different morphologies can be directed by agents, such as silk and other secondary chemicals, during reduction [37]. The minor presence of palladium hydride in the Pd–silk aerogel composites is likely due to hydrogen gas evolution during electrochemical reduction and the tendency of palladium to store hydrogen gas within its crystal lattice [38]. The palladium hydride peaks shift the position of the fitted palladium phase peaks slightly from their indexed positions. For instance, the (111) palladium fitted peak at 39.5° is shifted right relative to the indexed reference peak position at 39.0° likely due to peak convolution with the (101) palladium hydride peak. The change in nanocrystal structure can be associated with the shape-directing capabilities of the hydrogen evolution during reduction, which is observed as the visible peak shift in the XRD spectra [37]. The crystallite sizes, determined by using the Debeye–Scherrer formula, and the (111) peaks were 3.6 nm and 2.6 nm for silk–palladium and silk–platinum, respectively, and suggest that the nanoparticles observed in the SEM images in Figure 2 are polycrystalline. The XRD spectra for the HFIP–silk composite aerogel fiber with gold is shown in Figure S5a. The peaks observed for both the silk–palladium and silk–platinum composite aerogels at approximately 20.8° are attributed to the silk protein templates [39–41].

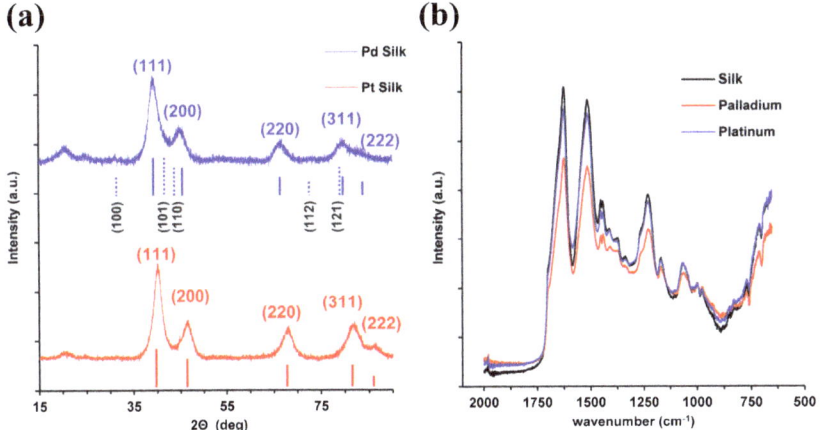

Figure 3. (a) X-ray diffraction spectra for silk palladium and platinum composite aerogels. The silk–palladium aerogel peaks are indexed to the Joint Committee on Powder Diffraction Standards (JCPDS) reference 01-087-0637 for palladium (blue lines), 01-073-0004 for palladium hydride (blue dashed lines; Miller indices labeled in gray). The silk–platinum aerogels are indexed to 00-004-0802 (red lines) for platinum. (b) Fourier transform infrared (FTIR) spectra for the silk, silk–palladium, and silk–platinum fiber aerogels.

3.4. FTIR Characterization

The secondary structure of the silk fibroin was completed by FTIR, examining the secondary structure in the Amide II and III band. The FTIR spectra for silk, silk–palladium, and silk–platinum are shown in Figure 3b [42]. The Amide I band is associated with 1600–1700 cm^{-1}, which shows the characteristic peak of beta-sheeted silk fibroin at 1625 cm^{-1} [20]. The alpha-helix structure (1658–1662 cm^{-1}) additionally shows a sharp peak associated with the shoulder near the Amide I band [20]. The FTIR spectra are characteristic of silk and are unchanged for the silk–palladium and silk–platinum aerogel fibers. After supercritical drying there is a high percentage of alpha-helix and beta-sheet content through the interaction at the molecular level. The same was found in the silk–gold aerogel fibers (Figure S5b). The high beta-sheet content observed is typical of HFIP–silk and has been shown in other studies where silk gels dried in ethanol or methanol are used for fracture fixation devices [27].

3.5. TGA Characterization

To characterize the mass composition of noble metals in the silk aerogels, thermogravimetric analysis (TGA) was performed with the results shown in Figure 4. When examined, the control silk, silk–palladium, and silk–platinum aerogel fibers showed an increase in the metal-to-silk weight ratio. Silk begins decomposition above 200 °C, but to ensure the complete degradation of the silk fibroin from the sample, analysis was conducted up to 1000 °C. The final mass percentage indicates the amount of noble metal to silk fibroin. The silk-only control sample shows 0% metal at 1000 °C [20,43,44], silk–palladium shows 10%, and silk–platinum indicates a mass percentage of 15%. The silk–gold aerogel fiber composites showed a higher metal mass percentage of 30%, which was observed in the morphological changes in the SEM images described above (Figure S6). The change in mass percentage of noble metal varied between each condition as stated above due to the electrostatic forces between the metal ions and silk fibroin. As stated above, the different pH values for 100 mM Na_2PdCl_4, K_2PtCl_4, or $HAuCl_4 \cdot 3H_2O$ cause different ionic interactions with the silk. Lower pH will deposit more metal ions, which, after reduction, shows a higher density of nanoparticles on the silk surface and is consistent with the mass percent differences observed with TGA.

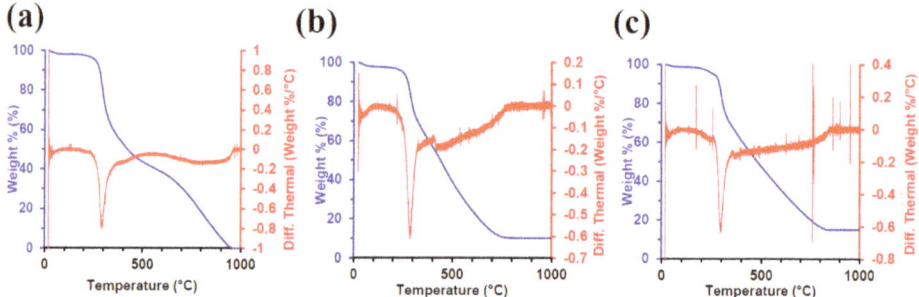

Figure 4. Thermogravimetric analysis (TGA) with differential thermal analysis (DTA) for the (**a**) silk aerogel, (**b**) silk–palladium aerogel, and (**c**) silk–platinum aerogel.

3.6. Porosity and Surface Area Characterization

Nitrogen gas adsorption isotherms were generated for the silk–noble metal aerogels prepared with 100 mM noble metals. Given the small mass of the silk aerogel composite fibers, bulk silk and silk composite aerogels were used to achieve nitrogen gas adsorption–desorption isotherms. The physisorption data shown in Figure 5 indicate type IV adsorption–desorption isotherms according to IUPAC classification standards, suggesting the presence of both mesoporous (2–50 nm) and macroporous (>50 nm) structures in all of the aerogel samples. The macropore features generally correlate with the pores observed in the SEM images in Figure 2 for the silk–palladium and silk–platinum composites. The mesoporous content suggests porosity within the silk phase of the composite aerogels not directly observed with SEM. After a sharp rise in adsorbed gas, no limiting adsorption plateau at high relative pressures (P/P_o) is observed. The maximum volume adsorbed at the highest relative pressure of P/P_o = 0.995 is 498, 482, and 254 cm^3/g for the silk, silk–palladium, and silk–platinum samples, respectively. All the sample isotherms exhibit type H3 hysteresis typical of mesoporous capillary condensation. The hysteresis loops close at higher relative pressures for the silk–palladium and silk–platinum composite aerogels compared with the silk only aerogels. This may be due to metal nanoparticles filling or blocking the smaller pores. This is supported by the pore frequency distributions seen in Figure 5b,d,e for silk, silk–palladium, and silk–platinum, respectively. The silk aerogels exhibit a high frequency of 3–4 nm pores, with a broad presence of mesopores up to 40 nm. The frequency of 3–4 nm pores decreases for the silk–palladium and silk–platinum aerogels with a commensurate increase in mesopores of diameters of 20–30 nm. The BET specific surface areas determined from desorption isotherms of the silk control, silk–palladium, and silk–platinum samples were 268, 170, and 72 m^2/g, respectively. The decrease in specific surface area from pure silk to silk–metal composite aerogels is consistent with nanoparticle pore blockage, which is consistent with SEM image analysis.

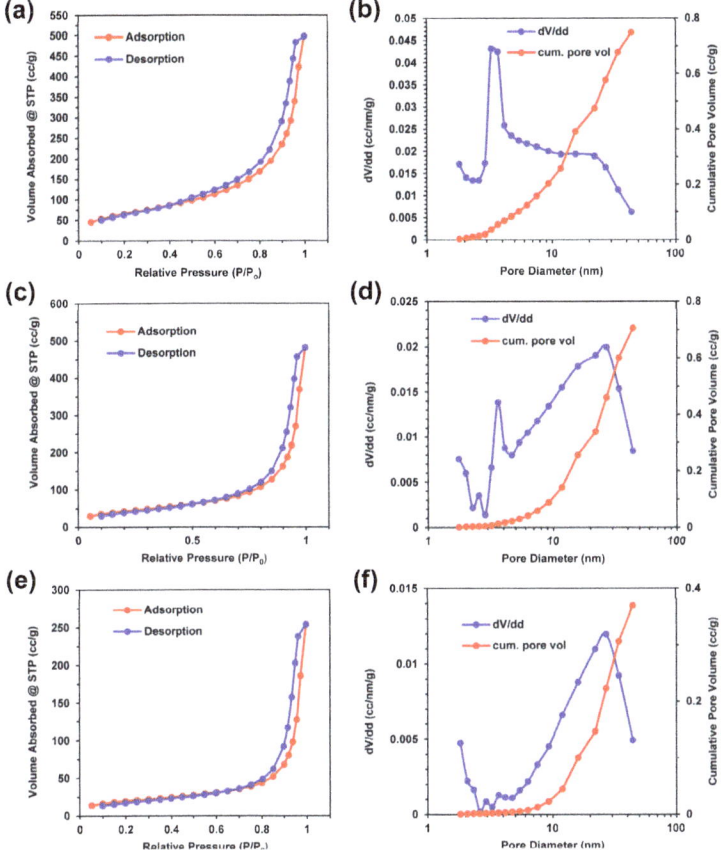

Figure 5. Nitrogen adsorption–desorption isotherms and pore size distribution with cumulative pore volume for the (**a**,**b**) silk aerogels, (**c**,**d**) palladium–silk aerogels, and (**e**,**f**) platinum–silk aerogels.

4. Conclusions

Here we have shown that HFIP–silk fibroin aerogel fibers can be utilized as a platform to anchor noble metal nanoparticles onto the surface of silk nanofibrils. The silk aerogels have a high surface area and are composed of different percentages of noble metals, which are based primarily on the nobility, not the concentration. This changes the mass percentage, porosity, surface area, and pore diameter. Compared with other biopolymers, which have acted as templates with noble metals, silk is much simpler to process and does not require any chemical crosslinking. Palladium, platinum, and gold nanoparticle growth have been shown on silk fibroin aerogel fibers. The molecular structure of silk is unchanged when anchored with different noble metals. These silk fibroin noble metal aerogels have potential utility in energy storage and conversion that can be used where degradable, flexible materials are required.

5. Patents

This work has been submitted with a preliminary patent and invention disclosure.

Supplementary Materials: The following are available online at http://www.mdpi.com/1996-1944/12/6/894/s1, Figure S1: Silk hydrogels equilibrated in 100 mM Na_2PdCl_4 and 100 mM K_2PtCl_6 reduced in 2 M $NaBH_4$; Figure S2: Silk hydrogels equilibrated in 100 mM $HAuCl_4$ reduced with 2 M DMAB; Figure S3: SEM images of the HFIP–silk noble metal fibers; Figure S4: SEM images of the HFIP–silk aerogels for silk–palladium, silk–platinum, and silk–gold; Figure S5: X-ray diffraction (XRD) and Fourier transform infrared (FTIR) absorption spectra for the silk–gold composite aerogels, Figure S6: TGA curves of the silk–gold aerogels; Figure S7: Nitrogen gas adsorption–desorption curves for the silk–gold aerogels.

Author Contributions: Conceptualization, A.N.M. and F.J.B.; Data curation, A.N.M., F.J.B. and C.K.N.; Formal analysis, A.N.M. and F.J.B.; Funding acquisition, A.N.M. and F.J.B.; Investigation, A.N.M., F.J.B., C.K.N., M.Y.R., J.W. and R.K.S.; Methodology, A.N.M. and F.J.B.; Project administration, A.N.M. and F.J.B.; Resources, A.N.M., F.J.B. and J.K.W.; Supervision, A.N.M. and F.J.B.; Validation, F.J.B.; Original draft preparation, A.N.M. and F.J.B.; Review and editing of manuscript, A.N.M., F.J.B., C.K.N., E.A.N., M.Y.R., J.W., R.K.S., K.W. and J.K.W.

Funding: This work was funded by a Faculty Development Research Fund grant from the United States Military Academy. A.N.M. received partial funding from the National Academy of Sciences through the National Research Council Davies Fellowship and through the Army Research Office (ARO).

Acknowledgments: The authors acknowledge Stephen Bartolucci at Watervliet Arsenal and Benet Laboratories for providing use of their SEM. The authors also acknowledge David Kaplan from Tufts University for the raw silk cocoons.

Conflicts of Interest: The authors declare no conflicts of interest.

References

1. Ulery, B.D.; Nair, L.S.; Laurencin, C.T. Biomedical Applications of Biodegradable Polymers. *J. Polym. Sci. B Polym. Phys.* **2011**, *49*, 832–864. [CrossRef] [PubMed]
2. Tan, J.; Saltzman, W.M. Biomaterials with Hierarchically Defined Micro- and Nanoscale Structure. *Biomaterials* **2004**, *25*, 3593–3601. [CrossRef] [PubMed]
3. Zhu, Z.H. Preparation and Characterization of Regenerated Bombyx Mori Silk Fibroin Fiber with High Strength. *eXPRESS Polym. Lett.* **2008**, *2*, 885–889. [CrossRef]
4. Burpo, F.J.; Mitropoulos, A.N.; Nagelli, E.A.; Rye, M.Y.; Palmer, J.L. Gelatin Biotemplated Platinum Aerogels. *MRS Adv.* **2018**, *3*, 2875–2880. [CrossRef]
5. Burpo, F.J.; Mitropoulos, A.N.; Nagelli, E.A.; Palmer, J.L.; Morris, L.A.; Ryu, M.Y.; Kenneth Wickiser, J. Cellulose Nanofiber Biotemplated Palladium Composite Aerogels. *Molecules* **2018**, *23*, 1405. [CrossRef] [PubMed]
6. Altman, G.H.; Diaz, F.; Jakuba, C.; Calabro, T.; Horan, R.L.; Chen, J.; Lu, H.; Richmond, J.; Kaplan, D.L. Silk-Based Biomaterials. *Biomaterials* **2003**, *24*, 401–416. [CrossRef]
7. Vepari, C.; Kaplan, D.L. Silk as a Biomaterial. *Prog. Polym. Sci.* **2007**, *32*, 991–1007. [CrossRef]
8. Lawrence, B.D.; Cronin-Golomb, M.; Georgakoudi, I.; Kaplan, D.L.; Omenetto, F.G. Bioactive Silk Protein Biomaterial Systems for Optical Devices. *Biomacromolecules* **2008**, *9*, 1214–1220. [CrossRef]
9. Lu, Q.; Hu, X.; Wang, X.; Kluge, J.A.; Lu, S.; Cebe, P.; Kaplan, D.L. Water-Insoluble Silk Films with Silk I Structure. *Acta Biomater.* **2010**, *6*, 1380–1387. [CrossRef]
10. Perry, H.; Gopinath, A.; Kaplan, D.L.; Dal Negro, L.; Omenetto, F.G. Nano- and Micropatterning of Optically Transparent, Mechanically Robust, Biocompatible Silk Fibroin Films. *Adv. Mater.* **2008**, *20*, 3070–3072. [CrossRef]
11. Jin, H.J.; Park, J.; Karageorgiou, V.; Kim, U.J.; Valluzzi, R.; Cebe, P.; Kaplan, D.L. Water-Stable Silk Films with Reduced β-Sheet Content. *Adv. Funct. Mater.* **2005**, *15*, 1241–1247. [CrossRef]
12. Wang, X.; Yucel, T.; Lu, Q.; Hu, X.; Kaplan, D.L. Silk Nanospheres and Microspheres from Silk/pva Blend Films for Drug Delivery. *Biomaterials* **2010**, *31*, 1025–1035. [CrossRef] [PubMed]
13. Kim, U.-J.; Park, J.; Li, C.; Jin, H.-J.; Valluzzi, R.; Kaplan, D.L. Structure and Properties of Silk Hydrogels. *Biomacromolecules* **2004**, *5*, 786–792. [CrossRef] [PubMed]
14. Floren, M.L.; Spilimbergo, S.; Motta, A.; Migliaresi, C. Carbon Dioxide Induced Silk Protein Gelation for Biomedical Applications. *Biomacromolecules* **2012**, *13*, 2060–2072. [CrossRef] [PubMed]
15. Mallepally, R.R.; Marin, M.A.; McHugh, M.A. CO2-Assisted Synthesis of Silk Fibroin Hydrogels and Aerogels. *Acta Biomater.* **2014**, *10*, 4419–4424. [CrossRef] [PubMed]
16. Marin, M.A.; Mallepally, R.R.; McHugh, M.A. Silk Fibroin Aerogels for Drug Delivery Applications. *J. Supercrit. Fluids* **2014**, *91*, 84–89. [CrossRef]
17. Tseng, P.; Napier, B.; Zhao, S.; Mitropoulos, A.N.; Applegate, M.B.; Marelli, B.; Kaplan, D.L.; Omenetto, F.G. Directed Assembly of Bio-Inspired Hierarchical Materials with Controlled Nanofibrillar Architectures. *Nat. Nanotechnol.* **2017**, *12*, 474–480. [CrossRef] [PubMed]

18. Zhao, C.; Yao, J.; Masuda, H.; Kishore, R.; Asakura, T. Structural Characterization and Artificial Fiber Formation of Bombyx Mori Silk Fibroin in Hexafluoro-Iso-Propanol Solvent System. *Biopolymers* **2003**, *69*, 253–259. [CrossRef] [PubMed]
19. Chen, X.; Shao, Z.; Knight, D.P.; Vollrath, F. Conformation Transition Kinetics of Bombyx Mori Silk Protein. *Proteins* **2007**, *68*, 223–231. [CrossRef]
20. Hu, X.; Kaplan, D.; Cebe, P. Determining Beta-Sheet Crystallinity in Fibrous Proteins by Thermal Analysis and Infrared Spectroscopy. *Macromolecules* **2006**, *39*, 6161–6170. [CrossRef]
21. Hirota, N.; Mizuno, K.; Goto, Y. Cooperative Alpha-Helix Formation of Beta-Lactoglobulin and Melittin Induced by Hexafluoroisopropanol. *Protein Sci.* **1997**, *6*, 416–421. [CrossRef] [PubMed]
22. Roccatano, D.; Fioroni, M.; Zacharias, M.; Colombo, G. Effect of Hexafluoroisopropanol Alcohol on the Structure of Melittin: A Molecular Dynamics Simulation Study. *Protein Sci.* **2005**, *14*, 2582–2589. [CrossRef] [PubMed]
23. Mandal, B.B.; Grinberg, A.; Seok Gil, E.; Panilaitis, B.; Kaplan, D.L. High-Strength Silk Protein Scaffolds for Bone Repair. *Proc. Natl. Acad. Sci. USA* **2012**, *109*, 7699–7704. [CrossRef] [PubMed]
24. Ling, S.; Qin, Z.; Li, C.; Huang, W.; Kaplan, D.L.; Buehler, M.J. Polymorphic Regenerated Silk Fibers Assembled through Bioinspired Spinning. *Nat. Commun.* **2017**, *8*, 1387. [CrossRef] [PubMed]
25. Zhang, W.; Ahluwalia, I.P.; Literman, R.; Kaplan, D.L.; Yelick, P.C. Human Dental Pulp Progenitor Cell Behavior on Aqueous and Hexafluoroisopropanol Based Silk Scaffolds. *J. Biomed. Mater. Res. Part A* **2011**, *97*, 414–422. [CrossRef] [PubMed]
26. Gil, E.S.; Kluge, J.A.; Rockwood, D.N.; Rajkhowa, R.; Wang, L.; Wang, X.; Kaplan, D.L. Mechanical Improvements to Reinforced Porous Silk Scaffolds. *J. Biomed. Mater. Res. Part A* **2011**, *99*, 16–28. [CrossRef] [PubMed]
27. Perrone, G.S.; Leisk, G.G.; Lo, T.J.; Moreau, J.E.; Haas, D.S.; Papenburg, B.J.; Golden, E.B.; Partlow, B.P.; Fox, S.E.; Ibrahim, A.M.S.; et al. The Use of Silk-Based Devices for Fracture Fixation. *Nat. Commun.* **2014**, *5*, 3385. [CrossRef] [PubMed]
28. Tsotsas, E.; Mujumdar, A.S. (Eds.) *Modern Drying Technology*; Wiley-VCH Verlag GmbH & Co. KGaA: Weinheim, Germany, 2011.
29. Williams, J.R.; Clifford, A.A.; Al-Saidi, S.H.R. Supercritical Fluids and Their Applications in Biotechnology and Related Areas. *Mol. Biotechnol.* **2002**, *22*, 263–286. [CrossRef]
30. Brunner, G. Applications of Supercritical Fluids. *Annu. Rev. Chem. Biomol. Eng.* **2010**, *1*, 321–342. [CrossRef] [PubMed]
31. Rockwood, D.N.; Preda, R.C.; Yücel, T.; Wang, X.; Lovett, M.L.; Kaplan, D.L. Materials Fabrication from Bombyx Mori Silk Fibroin. *Nat. Protoc.* **2011**, *6*, 1612–1631. [CrossRef]
32. Burpo, F.J.; Nagelli, E.A.; Morris, L.A.; McClure, J.P.; Ryu, M.Y.; Palmer, J.L. Direct Solution-Based Reduction Synthesis of Au, Pd, and Pt Aerogels. *J. Mater. Res.* **2017**, *32*, 4153–4165. [CrossRef]
33. Sing, K.S.W. Reporting Physisorption Data for Gas/solid Systems with Special Reference to the Determination of Surface Area and Porosity. *Pure Appl. Chem.* **1985**, *57*, 603. [CrossRef]
34. Brunauer, S.; Emmett, P.H.; Teller, E. Adsorption of Gases in Multimolecular Layers. *J. Am. Chem. Soc.* **1938**, *60*, 309–319. [CrossRef]
35. Barrett, E.P.; Joyner, L.G.; Halenda, P.P. The Determination of Pore Volume and Area Distributions in Porous Substances. I. Computations from Nitrogen Isotherms. *J. Am. Chem. Soc.* **1951**, *73*, 373–380. [CrossRef]
36. Jin, H.-J.; Kaplan, D.L. Mechanism of Silk Processing in Insects and Spiders. *Nature* **2003**, *424*, 1057–1061. [CrossRef] [PubMed]
37. Polavarapu, L.; Mourdikoudis, S.; Pastoriza-Santos, I.; Perez-Juste, J. Nanocrystal Engineering of Noble Metals and Metal Chalcogenides: Controlling the Morphology, Composition and Crystallinity. *CrystEngComm* **2015**, *17*, 2727–2762. [CrossRef]
38. Jewell, L.L.; Davis, B.H. Review of Absorption and Adsorption in the Hydrogen-Palladium System. *Appl. Catal. A Gen.* **2006**, *310*, 1–15. [CrossRef]
39. Zhang, H.; Li, L.L.; Dai, F.Y.; Zhang, H.H.; Ni, B.; Zhou, W.; Yang, X.; Wu, Y.Z. Preparation and Characterization of Silk Fibroin as a Biomaterial with Potential for Drug Delivery. *J. Transl. Med.* **2012**, *10*, 117. [CrossRef]
40. Lu, S.; Li, J.; Zhang, S.; Yin, Z.; Xing, T.; Kaplan, D.L. The Influence of the Hydrophilic-Lipophilic Environment on the Structure of Silk Fibroin Protein. *J. Mater. Chem. B* **2015**, *3*, 2599–2606. [CrossRef]

41. Wang, H.Y.; Zhang, Y.Q. Effect of Regeneration of Liquid Silk Fibroin on Its Structure and Characterization. *Soft Matter* **2013**, *9*, 138–145. [CrossRef]
42. Hu, X.; Kaplan, D.; Cebe, P. Dynamic Protein-Water Relationships during Beta-Sheet Formation. *Macromolecules* **2008**, *41*, 3939–3948. [CrossRef]
43. Lamoolphak, W.; De-Eknamkul, W.; Shotipruk, A. Hydrothermal Production and Characterization of Protein and Amino Acids from Silk Waste. *Bioresour. Technol.* **2008**, *99*, 7678–7685. [CrossRef] [PubMed]
44. Kang, K.Y.; Chun, B.S. Behavior of Hydrothermal Decomposition of Silk Fibroin to Amino Acids in near-Critical Water. *Korean J. Chem. Eng.* **2004**, *21*, 654–659. [CrossRef]

© 2019 by the authors. Licensee MDPI, Basel, Switzerland. This article is an open access article distributed under the terms and conditions of the Creative Commons Attribution (CC BY) license (http://creativecommons.org/licenses/by/4.0/).

Article

Roles of Silk Fibroin on Characteristics of Hyaluronic Acid/Silk Fibroin Hydrogels for Tissue Engineering of Nucleus Pulposus

Tze-Wen Chung [1,2,*], Weng-Pin Chen [3,4,*], Pei-Wen Tai [1,†], Hsin-Yu Lo [1] and Ting-Ya Wu [1]

1. Department of Biomedical Engineering, National Yang-Ming University, Taipei 11221, Taiwan; ylno02ekil@yahoo.com.tw (P.-W.T.); catfish19930628@gmail.com (H.-Y.L.); wuyaya0317@gmail.com (T.-Y.W.)
2. Center for Advanced Pharmaceutical Science and Drug Delivery, National Yang-Ming University, Taipei 11221, Taiwan
3. Department of Mechanical Engineering, National Taipei University of Technology, Taipei 10608, Taiwan
4. Additive Manufacturing Center for Mass Customization Production, National Taipei University of Technology, Taipei 10608, Taiwan
* Correspondence: twchung@ym.edu.tw (T.-W.C.); wpchen@ntut.edu.tw (W.-P.C.); Tel.: +886-2-28267019 (T.-W.C.)
† Co-first author.

Received: 28 February 2020; Accepted: 11 June 2020; Published: 17 June 2020

Abstract: Silk fibroin (SF) and hyaluronic acid (HA) were crosslinked by horseradish peroxidase (HRP)/H_2O_2, and 1,4-Butanediol di-glycidyl ether (BDDE), respectively, to produce HA/SF-IPN (interpenetration network) (HS-IPN) hydrogels. HS-IPN hydrogels consisted of a SF strain with a high content of tyrosine (e.g., strain A) increased viscoelastic modules compared with those with low contents (e.g., strain B and C). Increasing the quantities of SF in HS-IPN hydrogels (e.g., HS7-IPN hydrogels with weight ratio of HA/SF, 5:7) increased viscoelastic modules of the hydrogels. In addition, the mean pores size of scaffolds of the model hydrogels were around 38.96 ± 5.05 µm which was between those of scaffolds H and S hydrogels. Since the viscoelastic modulus of the HS7-IPN hydrogel were similar to those of human nucleus pulposus (NP), it was chosen as the model hydrogel for examining the differentiation of human bone marrow-derived mesenchymal stem cell (hBMSC) to NP. The differentiation of hBMSC induced by transforming growth factor β3 (TGF-β3) in the model hydrogels to NP cells for 7 d significantly enhanced the expressions of glycosaminoglycan (GAG) and collagen type II, and gene expressions of aggrecan and collagen type II while decreased collagen type I compared with those in cultural wells. In summary, the model hydrogels consisted of SF of strain A, and high concentrations of SF showed the highest viscoelastic modulus than those of others produced in this study, and the model hydrogels promoted the differentiation of hBMSC to NP cells.

Keywords: silk fibroin; hyaluronic acid; Tyrosine; viscoelastic modulus of HS-IPN hydrogels; hBMSC differentiations; nucleus pulposus

---

1. Introduction

Hydrogels can be produced by crosslinking polymers to form interpenetration network (IPN) with varying mechanical properties [1,2] for tissue engineering such as cardiac tissue repairs [3], controlling the fates of stem cells [1], and drug delivery, etc. [4]. Since hydrogels are highly permeable to nutrients and water-soluble metabolites, they can support cell growth and proliferation which are suitable for tissue engineering (TE). Hydrogels for TE usually consist of synthetic polymers, such as polyurethane and polyvinyl alcohol (PVA) [5,6], or natural polymers, such as HA and collagen [7,8].

HA is a natural glycosaminoglycan with carboxylic groups; it is an important component of the extracellular matrices (ECM) in various tissues and play important roles in cell proliferation and migration [9]. For instance, the interactions of HA in a cardiac patch and CD44 of BMSC enhanced cardiac differentiations of BMSC in both cardiac gene and proteins expressions [10]. In addition, various methods to prepare HA-based hydrogels including oxidized-HA or methacrylated-HA have been investigated for TE of ECM of NP [11–13], respectively.

Various of SF-based membranes, scaffolds or hydrogels have been extensively studied for the applications of TE because of its favorable biological responses, such as weak antigenic effects and inflammatory responses in-vivo [9,14–16]. For example, SF/HA patches laden with hBMSC promoted cardiac repair in a rat myocardial infarction (MI) model [10,16]. Developing suitable mechanical properties for hydrogels is also important to enhance cell proliferations and hBMSC differentiations for using various TE [17,18]. In this regard, the crosslinking tyrosine in SF to produce di-tyrosine bonds by HRP/H_2O_2 enzymatic reactions produced silk elastomers with stiffness that are varied from 0.2 to 10 kPa [17]. However, the stiffness of the aforementioned SF hydrogels was not suitable for the needs of some tissues such as human NP. Although the influences of molecular weights of SF on mechanical properties of SF hydrogels have been reported [17], the influence of total tyrosine contents in SF on the mechanical properties of SF-based hydrogels has not been investigated.

The intervertebral disc (IVD) absorbs shocks by transferring and dissipating loads to the ECM within the superior and inferior discs. IVD degeneration generally causes lower back pain, which is a common health problem. [7,18]. Currently, clinical treatments, such as spinal fusion and partial or total disc replacement cannot fully restore or maintain IVD structures and functions [7,18,19]. Since disc degeneration originates in nucleus pulposus (NP) regions, tissue engineering NP, that may mimic the structure of native NP tissue and possibly fully restore the functionality of healthy IVD discs, is great needed [18,19].

Various biomaterials have been investigated for TE of NP. They are: A. carbohydrate polymers including HA, dextran, chitosan and carboxymethylcellulose (CMC) to produce various hydrogels such as oxidized-HA or methacrylated-HA [11–13], dextran-chitosan-teleostean and methacrylated CMC hydrogels, respectively [7,18,19]; B. proteins including Type II collagen hydrogels, and laminin-based hydrogels, respectively [20,21]; C. hybrid of carbohydrate polymers and proteins, such as crosslinked oxidized HA/gelatin [12] and Type II collagen with a low molecular weight of HA [8,22]. However, the mechanical properties of some of the aforementioned biomaterials were not suitable for TE of NP, which need to be further processed [11,20,21]. For instance, the elastic modulus of the oxidized-HA hydrogels, using adipic acid di-hydrazide (ADH), are much lower than those of the native ECM of the NP [11]. Although hydrogels produced by oxidized HA-gelatin using ADH could improve the elastic modulus of the hydrogels (~11 kPa), the controlling the chemical reactions of HA/gelatin by ADH might not be easily carried out [12]. Recently, dextran-chitosan-teleostean triple-interpenetrating network and methacrylatedC MC hydrogels has been examined in a goat model to support the mechanical functions of degenerative NP [18,19]. However, chitosan and CMC had not yet been approved by the FDA for use in internal organs.

Although the bioactivities of SF or HA/SF patches for cardiac repairs have been shown in-vitro, and in-vivo, respectively [10,16], the mechanical properties for HS-IPN hydrogels, which consisted of various amounts of tyrosine in SF, in terms of vary strains of SF in this study, and the weight ratios of SF to HA have not been investigated. Moreover, the potential of differentiations of induced hBMSC laden-HS-IPN hydrogels to NP cells has not been examined. Although using NHS/EDC to crosslink HA/SF to facilely prepare HA/SF hydrogels was recently reported [23], it was one step of our processes to prepare hydrogels. However, they did not examine the rheological properties of hydrogels. Using HRP/H_2O_2 to crosslink tyramine-substitute HA and 2% of SF to produce hydrogels with varying HA contents has also been reported by Raia et al. [24]. HA and SF-IPN hydrogels produced by them are configured by di-tyramine bonds in HA crosslinked network, di-tyrosine bonds in SF crosslinked network and tyramine-tyrosine bonds in HA/SF-IPN. The tyramine-substitute HA needed to be complexly and chemically synthesized for the research study, which was not a commercially available biomaterial.

To produce HS-IPN hydrogels, HA-crosslinked network hydrogels were first produced by using BDDE solutions to crosslink HA polymers (Scheme 1). The influences of varying strains of SF obtained from *B. mori* cocoons and, at a fixed quantities of HA, varying amounts of SF (expressed in wt. ratios of 5/1~5/7 for HA/SF, respectively) were crosslinked using HRP/H_2O_2 reactions to produce HS-IPN hydrogels. To further increase in-vitro stability of hydrogels, the carboxyl acids of HA in HS-IPN hydrogels were crosslinked with amine groups of added polyethyleneimine (PEI) and HS-IPN using *N*-(3-Dimethyl-amino-propyl)-*N'*-ethylcarbodiimide hydrochloride)/*N*-Hydroxy-succinimide (EDC/NHS) reagents (Scheme 1). The differentiation of hBMSC laden on HS-IPN hydrogels, induced using TGF-β3 to NP cells were examined for TE of NP.

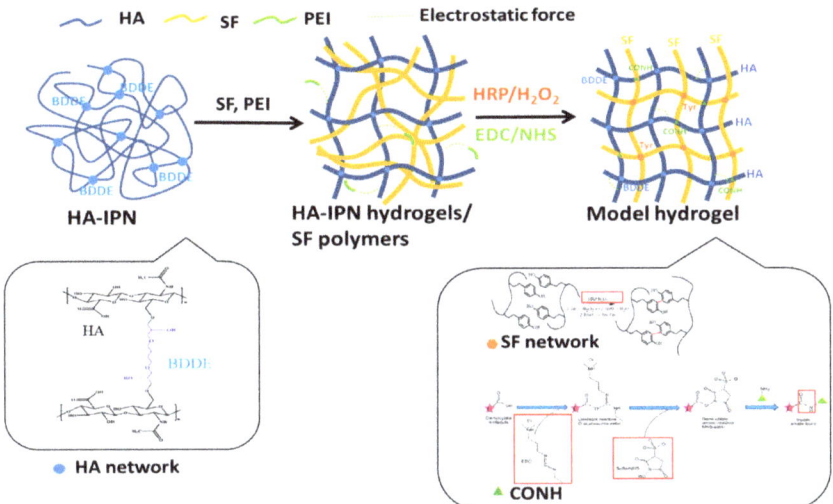

Scheme 1. The schematic procedures to develop HS-IPN hydrogel; HA polymer chains were first crosslinked by BDDE reagent for several hours to produce HA crosslinked network hydrogel (blue dots); SF and a small quantity of PEI (polyethyleneimine) were added into the gel and well blended. SF polymer chains were then crosslinked by HRP/H_2O_2 enzymatic reactions in HA crosslinked network hydrogel to produce large quantities of di-tyrosine bonds, make SF crosslinked hydrogel, (brown dots or Tyr in the model hydrogels) and yield HS-IPN hydrogel. HS-IPN hydrogel and PEI were further crosslinked using EDC/NHS reagents to produce many amide bonds to stabilize the hydrogels.

2. Materials and Methods

2.1. Fabrication of Interpenetrating Network HA-SF Hydrogels

Various strains, strain A, B and C herein, of *B. mori* cocoons were the products of artificially cross breeding specific strains of *B. mori* to produce varying toughness of SF that might vary amino acid configurations of SF, including tyrosine contents. Each strain of *B. mori* was well-bred to control the quality of SF, including low variations in amino acid configurations of SF and its cocoon was gifted from MDARES (Miaoli Agricultural Research and Extension Station, Council of Agriculture, Executive Yuan, Miaoli, Taiwan). In addition, the stain A of SF was used in this study without being further specified.

Solution of SF (MW ~ 185 kDa) of strain A was prepared as described in early reports, such as degumming, de-solving in 9.3 M LiBr and removal of Li^+ using dialysis with DI water that were published by the authors' laboratory [10,16]. Briefly, silk cocoons were boiled in 0.02 M Na_2CO for 0.5 h and then rinsed thoroughly in D.I. water to extract the glue-like sericin proteins from silk fibroin, degumming procedures. The extracted SFs were then dissolved in 9.3 M LiBr solution at 60 °C for 4 h to yield a 20% (w/v) solution, which was then dialyzed against D.I. water using a dialysis membrane

(MWCO 6000) (Spectra/Por 1, Repligen Co., Waltham, MA, USA) at room temperature to remove salt for 48 h [10,16].

The MW of SF of strain A, B and C were determined by a SDS-PAGE method (Vertical Electrophoresis System, XCell SureLock™ Mini-Cell, Invitrogen, Waltham, MA, USA) with 150 V for 60 min. The bands of samples and molecular weight ladder (Himark Pre-Stained, Invitrogen, Waltham, MA, USA) were stained by a Coomassie Brilliant Blue R-250 (Sigma Corp., St. Louis, MO, USA). The images of the bands of the samples and standard molecular weight ladder were analyzed by an Image J (Windows version, Java 1.6.0-45, 32 bit mode, NIH, Bethesda, MD, USA). For amino acids analysis, 0.4 mg of dried SF samples (e.g., strand A) were hydrolyzed by 6 N HCL at 115 CC for 24 h. The analysis of amino acids of various samples were conducted by an amino acid analyzer (Hitachi L-8900, Tokyo, Japan).

To prepare HA hydrogel (H gel), HA (MW: 200 kDa, Lifecore Biomedical Inc., Chaska, MN, USA) was dissolved in NaOH at a concentration of 10%. 1,4-Butanediol diglycidyl ether (BDDE, Sigma Corp., St. Louis, MO, USA) was used for cross-linking reactions that was conducted at 37 °C for 6–8 h. The pH value of the solution was then adjusted to about 7.0 by adding HCl solution to terminate the crosslinking reaction [25,26]. After the HA-BDDE crosslinking reaction had been terminated, 0.15 mM PEI (polyethyleneimine, Mw. 25 KDa, Sigma, St. Louis, MO, USA), an appropriate amount of H_2O_2 (about 0.15 mM) and horseradish peroxidase (50 U/mL) (HRP, Sigma, St. Louis, MO, USA) and 5% SF solution were added to the H gels and well blended. SF in the aforementioned solutions were crosslinked and gradually turned to form the second interpenetration network, SF IPN-gels, because large quantities of di-tyrosine bonds of SF were produced by the reactions of HRP/H_2O_2, the enzyme reactions, in H gels for about 1 h at 37 °C [27].

To prepare SF hydrogel (S gel), an appropriate amount of H_2O_2 (about 0.15 mM) and horseradish peroxidase (50 U/mL) (HRP, Sigma, St. Louis, MO, USA) were added to 5% SF solution SF of strain A to induce enzymatic reaction for 1 h at 37 °C [27]. HS-IPN hydrogels were produced by blending the aforementioned H gels and S gels (Scheme 1).

To stabilize the HS-IPN hydrogels, 0.15 mM PEI in the hydrogels was further crosslinked by EDC/NHS (1:1) (N-(3-Dimethyl-amino-propyl)-N'-ethylcarbodiimide hydrochloride, $C_8H_{17}N_3$ HCL, EDC), Sigma Corp., St. Louis, MO, USA)/(N-Hydroxy-succinimide, $C_4H_5NO_3$, NHS), Fluka Chemical Corp., Rochester, NY, USA) for around 30 min. at temperature lower than 20 °C to produce amide bonds of carboxyl groups of HA with amine groups in PEI and in SF, respectively (Scheme 1). In addition, to examine the influences of varying ratios of SF on viscoelastic properties of HS-IPN hydrogels, the weight ratios of HA (4% at final) and SF in the hydrogels were changed to 5:1, 5:3, and 5:7 (e.g., HS1-IPN, HS3-IPN, HS7-IPN (namely, the model hydrogels)), respectively.

2.2. Characterizations of HS-IPN Hydrogels

2.2.1. Attenuated Total Reflectance-Fourier Transform Infrared (ATR-FTIR) Spectra of the Hydrogels

After various HS-IPN hydrogels were freeze-dried at −50 °C, their transmission spectra of the samples were examined using an ATR-FTIR instrument (IRAffinity-1, Shimadzu Co, Kyoto, Japan) with a resolution of 4 cm^{-1} at wave numbers 400–4000 cm^{-1} [10]. Detailed procedures for the measurements can be referred elsewhere. The spectra were analyzed using the built-in standard software package (IRAffinity-1, Shimadzu Co, Kyoto, Japan) [10].

2.2.2. Swelling Ratios of H, S, Crosslinked H/PEI and the Model Hydrogels

To study the swelling ratios of various gels, the samples were placed in phosphate saline (PBS) allowed for unrestricted deformation swelling test. H/PEI crosslinking hydrogels were fabricated after H gels were produced, followed by crosslinked 0.015 mM PEI in the gels by EDC/NHS crosslinking reagent for 30 min. The purpose of producing H/PEI crosslinking hydrogels was to examine swelling property of the gels after amide bonds formation between the carboxyl groups in HA gels and amine groups of PEI. The weights of hydrogels were measured at time of 0, 0.5, 1, 4, 24 and 48 h, W_t, after they were immersed into PBS for the aforementioned time. Their surfaces were gently wiped before they were weighted. The swelling ratio of a hydrogel was calculated by the following equation while W_{net} was the weight of the hydrogel before immersed into PBS:

$$\text{Swelling ratio (\%)} = \frac{W_t - W_{net}}{W_{net}} \times 100\% \tag{1}$$

2.2.3. Morphology of HS-IPN Hydrogels

To observe the morphology and the pore structure of HS-IPN hydrogels, all hydrogels were frozen at −50 °C and then at liquid nitrogen, further dried in vacuum for several days, so-called freeze-dry technique, to produce scaffolds. They were cut by a surgical knife to obtain the cross-section for observing the pore size distributions and characterizing the network in the scaffolds. All samples were coated with Pt and imaged with scanning electron microscopy (SEM, JSM-7600F, JEOL, Tokyo, Japan) at an accelerating voltage of 3–10 kV. The pore sizes of gel were calculated by ImagePro software (IPP7.0, Media Cybernetics, Rockville, MD, USA).

2.3. Rheological Studies of Vary Compositions of HS-IPN Hydrogels

The viscoelastic properties of various hydrogels were determined using a rheometer with an oscillatory mode (Discovery-HR1, TA Instruments, New Castle, DE, USA) at 37 °C. Oscillation frequency sweep tests using a parallel-plate sensor with a gap of 1 mm and loaded by 0.1 mL of HS-IPN hydrogels were carried out by the rheometer. The elastic modulus (G′) and viscous modulus (G″), phase shift angle (δ) and complex modulus (|G*|) of vary compositions for HS-IPN hydrogels were determined at a fixed strain of 0.01 rad with varying angular frequencies of 1–100 rad/s, as reported in other studies [11,12,28].

Compressive Modules of the Model Hydrogels

An MTS Model 858 Bionix Test System (MTS Corp., Eden Prairie, MN, USA) with a 5 kg load cell was used to perform confined compression tests on the porcine NP and l HS7-IPN hydrogels (or the model hydrogels). An acrylic cylinder with a diameter of 10 mm and a height of 30 mm was made and used as the indenter. A hollow, cylindrical acrylic container with an outer diameter of 30 mm, an inner diameter of 10 mm, and a height of 20 mm, was use in the confined compression test. A cavity with a diameter of 25.5 mm and a depth of 2.5 mm was cut into the bottom of the container and a porous plate was placed at the bottom of it. Porcine NP and hydrogel specimens were prepared and used to fill the acrylic container to a height of 5 mm. The confined compression tests were then carried out with five specimens for each material. A preload of 5 N (63.7 kPa) was applied to the indenter for 10 min to obtain a pressure balance on each specimen. In each confined compression test, following the preload phase, the indenter was moved at 0.1 mm/min until a final strain of 5% for the model hydrogels and the porcine NP while that of 15% (large strain compression) performed for the model hydrogels was mainly used for the comparisons with other studies. The load-displacement curve that was thus obtained for each specimen was then converted to a stress–strain (σ–ε) curve. The elastic modulus E was calculated as E = σ/ε for the linear portion of the stress-strain curve.

2.4. Cytotoxicity of the Model Hydrogels

To evaluate the biocompatibility of the model hydrogels, MTT (3-(4,5-dimethylthiazol-2-yl)-2,5-diphenyl-tetrazolium bromide), Sigma–Aldrich, St. Louis, MO, USA) assays were performed to test the cytotoxicity of vary ratios of extraction mediums, taken from supernatants, which was obtained by incubation the hydrogels with L929 fibroblasts for 24 h according to the guidelines of international standard organization (ISO) 10993-5 [29]. L929 fibroblasts were purchased from the Bio-resource Collection and Research Center (BCRC, Hsin-Chu, Taiwan) and cultured in Dulbecco's modified

Eagle's medium that contained 10% horse serum at 37 °C in a 5% CO2 incubator. After L929 cells were cultured with vary volume ratios of extraction mediums to cultural medium (e.g., 1/8, 1/4, 1/2 and 1.0 (or extraction medium only), namely dilution ratios) for 24 h, MTT assay to the cells were carried out to examine the cell viabilities as early report of this lab [29].

2.5. Differentiations of Induced hBMSC to NP in the Model Hydrogels

To evaluate the potential of the model hydrogels for NP regeneration, human bone marrow-derived mesenchymal stem cell (hBMSC) from passage 6~8 were used according to our previous study [10]. The density of hBMSC of 1×10^7 cell/mL was seeded on the surface layer of a 1.0 cm^2 the model hydrogels to produce the cell-laden hydrogels. The viability of hBMSC in the model hydrogels after 3 d of cultivation were stained with Live/Dead or Viability/Cytotoxicity kit (Invitrogen, Waltham, MA, USA) according to the manual instructions. Live/Dead stain of the cells was observed by a laser confocal scanning microscopy (LCSM) (Olympus FV1000, Olympus Corp., Tokyo Japan) [10].

To induce the chondrogenesis of hBMSC in the model hydrogels, after the cell proliferations in the hydrogels for 3 d, the culture medium was changed to chondrogenic differentiation medium containing TGF-β3 which was purchased from Lonza Corp. (Lonza, Gaithersburg, MD, USA). The differentiation medium was changed every two days, and followed by culture for 7 d and 14 d, respectively, for chondrogenesis study of hBMSC.

2.5.1. Immuno-Histochemical (IHC) Analysis of Specific Protein Expressions of the Differentiations of Induced hBMSC to NP

IHC analysis of specific protein expressions the differentiations of hBMSC to NP after induced by TGF-β3 was carried out at 7 d and 14 d. All samples were washed using PBS and fixed by 4% paraformaldehyde. After the fixation, they were moved from the cultural wells and then embedded in paraffin and sectioned into sections with a thickness of 5 µm. The sections were then stained with Alcian blue for examining the depositions of glycosaminoglycan (GAG) and collagen type II, respectively. Cell-free hydrogel was stained as a negative control.

2.5.2. Real-Time PCR for Specific Gene Expressions of the Differentiations of Induced hBMSC to NP

To determine the specific gene expressions of the differentiations of induced hBMSC to NP in the model hydrogels, the ECM-related gene expression, including Aggrecan (AGN), Collagen type I (Col I) and Collagen type II (Col II) were assessed by real-time PCR (StepOnePlus™ Real-Time PCR System, Applied Biosystems, Foster, CA, USA) at 7 d and 14 d, according to other studies [11,12,19]. GAPDH was used as the internal control. The relative gene expression was calculated as $2^{-\Delta\Delta C_t}$.

2.6. Statistics

All calculations are made using SigmaStat statistical software (Jandel Science, San Rafael, CA, USA) [10]. Statistical significance in the Student t-test corresponded to a confidence level of 95%. Data presented are mean ± SD from at least triplicate measurements. Differences were considered statistically significant at $p < 0.05$.

3. Results and Discussion

3.1. Fabricating Fluidity HA Hydrogels by Adjusting Parameters of HA, and Reactions Conditions of BDDE Crosslinking Reactions

HA hydrogels have high water contents because HA contents a large amounts of carboxyl groups and have been used for tissue repairs and in drug delivery systems [11]. However, without chemical modifications such as crosslinking of HA hydrogels, the gels would be easily disassembled in aqueous environment and then lost their mechanical properties which would usually deviate from those of human tissues including NP. Hydrogels for tissue repairs, including IVD repair, crosslinking HA to produce the HA network hydrogel has been widely prepared by oxidized HA or methacrylated HA, in order to improve and sustain the mechanical properties of HA [12–14]. However, it involves complex and delicate chemical reactions. Alternatively, HA network hydrogel was widely fabricated by crosslinking hydroxyl groups of N-acetyl-D-glucosamine (NAG) in HA using the epoxide groups of BDDE at a high pH condition (pH >11). In this study, HA network hydrogels were produced as the aforementioned method with modifying reaction parameters (Scheme 1). For the reactions in which HA was crosslinked by BDDE, several parameters such as the MW and concentrations of HA, the concentrations of BDDE, and the reaction time, would influence various properties of HA hydrogels [2,30]. For example, the uses of various concentrations of BDDE (0.01~20%) under alkaline conditions to crosslink various concentrations of HA with high molecular weight (2.65~10%, MW >10^3 kDa) to produce HA crosslinked hydrogels have been extensively investigated [30]. Here, a low concentration (~2.5%) and a low MW (~200 kDa) of HA was crosslinked using around 2.0% BDDE for 5–7 h to produce HA crosslinked hydrogels (Scheme 1), which had low G′ and G″ (e.g., 0.24 ± 0.092 kPa and 0.09 ± 0.005 kPa, $n = 3$, respectively) with phase angle, δ, 21.4°, indicating that viscoelastic HA crosslinked hydrogels were high fluidity. Hence, HA crosslinked hydrogels could be mixed well with varying amounts of SF for producing HS-IPN hydrogels.

Using HRP/H_2O_2 Reactions to Crosslink SF and Producing HS-IPN Hydrogels

Although the de-sericin process for SF polymers affects its molecular weight [31], the de-sericin procedures herein were well controlled which ensured the molecular weight of SF was approximately 185 kDa, as determined by SDS-PAGE (data not shown). Since high fluidity of HA crosslinked hydrogels could be well mixed with SF, using HRP/H_2O_2 enzymatic reactions to crosslink Tyr in SF within the hydrogels could homogenously take place to produce SF-IPN hydrogels and, consequently, produce HS-IPN hydrogels (Scheme 1). To further stabilize the HS-IPN hydrogels, 0.15 mM of PEI added into the hydrogels was further crosslinked by EDC/NHS for about 30 min. to produce amide bonds of carboxyl groups of HA with amine groups in PEI and in SF in the hydrogels, respectively. (Scheme 1).

According to the results of rheological study (in Section 3.3), the viscoelastic properties of HS7-IPN hydrogels were similar to matrix for human NP. The HS7-IPN hydrogels were chosen as a model hydrogel for inducing hBMSC to differentiate to NP cells or NP tissue engineering in this study.

3.2. Characterizations of HS-IPN Hydrogels

3.2.1. ATR-FTIR Spectra of HA, SF, and the Model Hydrogels Consisted of Varying Strains of SF

The model hydrogels were characterized by spectra using an ATR-FTIR spectrophotometer (Figure 1). In the Figure 1, the transmission spectra for the peaks of carboxyl groups of H gels such as 1600 and 1402 cm^{-1} were about the same as those of HA polymers. Moreover, the peak of amide II of H gels was 1556 cm^{-1}, which was similar to that of HA polymers. The transmission spectra for the peaks of amide I, II and III groups of S gels were 1640, 1511 and 1230 cm^{-1} that were about the same as SF polymers. The transmission spectra from HS-IPN gels produced from different strains of SF (e.g., strain A, B and C) contained several characteristic peaks as those of H gels and S gels with minor variations. For instance, the peaks of amide II of HS-IPN gels shifted from 1556 cm^{-1} of H gels to 1528, 1522 and 1525 cm^{-1} for strain A, B and C, respectively (Figure 1). The presences of the peak of amide I of HS-IPN hydrogels only minor shifted from 1640 cm^{-1} to 1635 cm^{-1} of S gels, and there was no difference among the model hydrogels consisted of varying strains of SF. Although the quantities of functional groups for varying weight ratios of HA/SF of the HS-IPN hydrogels were different such as HS1-IPN and the model hydrogels, the ATR-FTIR spectra for those hydrogels were similar to those of the model hydrogels and not able to characterize their differences (Figure 1). Hence, other characterizations including rheological properties of the hydrogels needed to be carried out to determine the differences among the model hydrogels consisted of varying strains of SF.

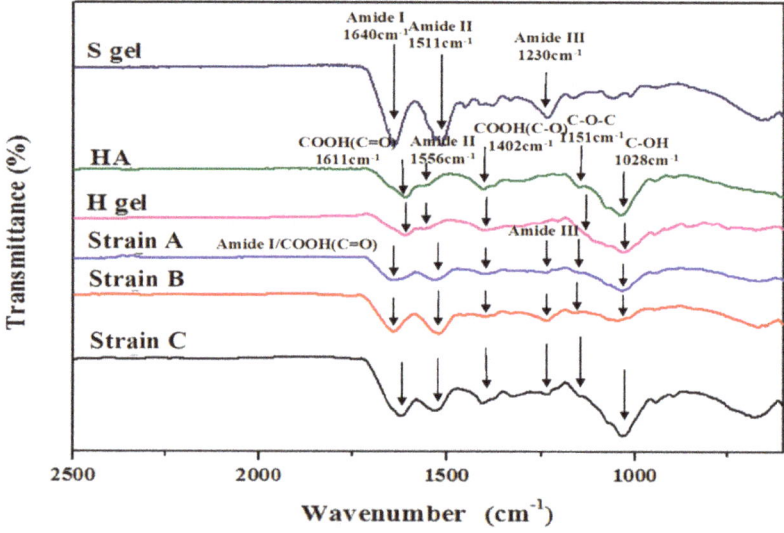

Figure 1. ATR-FTIR transmission spectra of varying functional groups of SF, HA polymers, and H, S and HS7-IPN hydrogels. The characteristic peaks of H hydrogels such as carboxyl groups (COOH), and S hydrogels such as amide I, II and III, respectively, were found in HS-IPN hydrogels without varying by the SF of strains of A, B and C.

3.2.2. The Pore Structures of Scaffolds and Swelling Ratios for H, S and the Model Hydrogels

To examine the pore structures of scaffolds for H, S and the model hydrogels, SEM micrographs of the scaffolds were carried out and presented (Figure 2). The mean pore sizes in H and S scaffolds were approximately 49.36 ± 15.04 μm, and 13.40 ± 1.25 μm, respectively, ($n = 3$). The mean pore size of the scaffolds of S gels was significantly smaller ($p < 0.01$, $n = 3$) than that for the scaffold of H gels (Figure 2). Interestingly, the mean pore size of the scaffolds of the model hydrogels was around 38.96 ± 5.05 μm ($n = 3$) which was between those H and S scaffolds. (Figure 2). In addition, the pore size of the scaffolds of the model hydrogels was suitable for the proliferations of cells, including hBMSC [7].

The swelling ratios of H, S, H/PEI crosslinking and the model hydrogels in PBS were examined and shown (Figure 3). Since H gels were produced by highly hydrophilic polymers, the swelling ratios of the hydrogels fast increased within 12 h till 550% and then slowly increased up to ~700% at 48 h which was significantly higher than other hydrogels ($p < 0.01$, $n = 3$). Interestingly, after the first 6 h of fast swelling stage, the swelling ratios for the model and H/PEI crosslinked hydrogels were in plateau regions (e.g., 185.9 ± 24.4%, $n = 3$) until the end of study, 48 h. However, after swelling in the first 4 h, the swelling ratios for S hydrogels in PBS were about 24.9 ± 0.9% till 48 h, which were significantly lower than others ($p < 0.01$, $n = 3$) (Figure 3). The swelling ratios of the model hydrogels were very close to those of H/PEI crosslinked hydrogels, and the values were between those for H and S gels [32]. Notably, H/PEI crosslinked hydrogels highly reduced the swelling ratios in PBS compared with those for H gels, indicating that EDC/NHS reactions among HA and PEI in hydrogels would effectively crosslink carboxylic and amine groups to produce the amide bonds within the gels which resulted in significant decreases of the interactions among carboxylic groups of H gels and ambient H_2O (Figure 3). Since EDC/NHS reactions were also carried out for preparing the model hydrogels as those for H/PEI crosslinked hydrogels, the swelling ratios for the model ones were similar to those of those for the later gels although the model hydrogels consisted of SF-IPN. In addition, the swelling ratios for other HS-IPN hydrogels (e.g., HS3-IPN hydrogels) would be similar to those for the model ones.

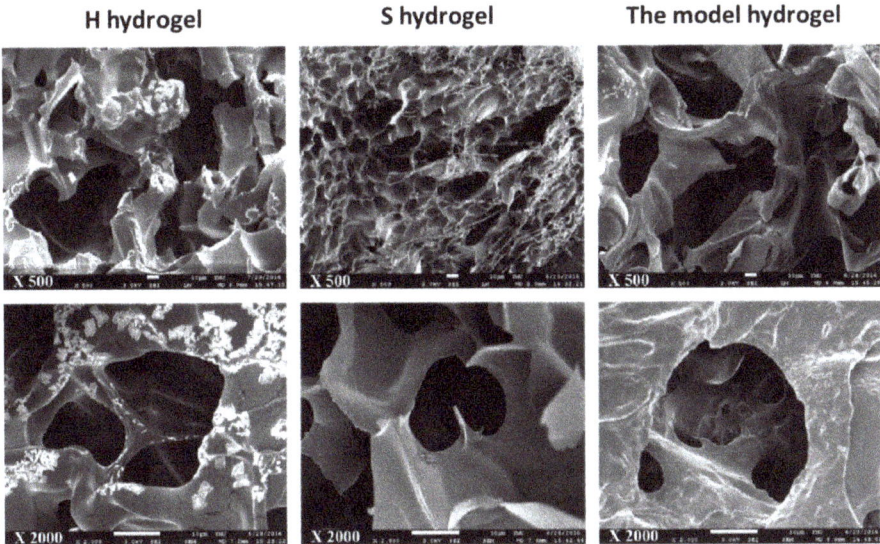

Figure 2. SEM micrographs of H, S, and HS-IPN hydrogels. The pore size of the model hydrogels was 38.96 ± 5.05 μm, (n = 3), in between those of H and S hydrogels.

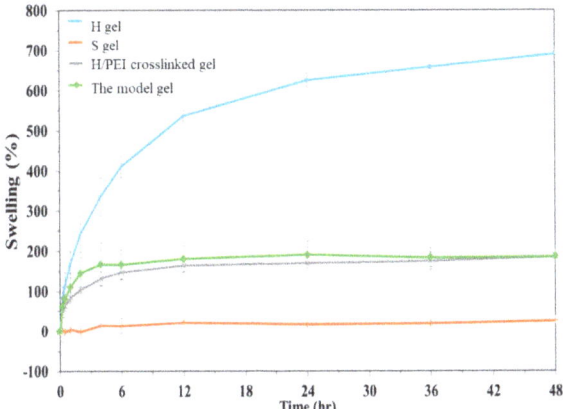

Figure 3. The swelling ratios for H, S, H/PEI crosslinked and the model hydrogels in PBS were shown. The swelling ratios for the H gels were significantly higher than others (p < 0.01, n = 3), while those for S gels were significantly lower than others (p < 0.01, n = 3). The swelling ratios of the model hydrogels in PBS were similar to those for crosslinking H/PEI crosslinked hydrogels between those of H and S hydrogels (Data are mean ± SD, n = 3).

Notably, the results or data of HS-IPN gels at slow swelling ratio stages were similar to human NP, revealing that they may be suitable for use in TE of NP [7,28]. Moreover, the stability of the model hydrogels in PBS solution was examined by observation and measuring weight loss of the hydrogels after they were immersed into PBS solution for three weeks. The morphology of the model hydrogels was intact without disintegration by observation while approximately 4% of weight loss was found compared to that of the original hydrogels to be immersed in PBS. In addition, the samples became more fluidity towards the end of the four weeks. Notably, the results of HS-IPN gels at slow swelling ratio stages were similar to human NP, revealing that they may be suitable for use in TE of NP [28].

3.3. Varying Strains of SF Influenced the Rheological Properties of HS-IPN Hydrogels

Viscoelastic flow parameters are important mechanical properties to characterize hydrogels for applications of bio-fluids such as lubricant in joints or NP. To determine, G', G" and δ values of HS-IPN hydrogels, they are determined using a parallel-plate rheometer operated in an oscillatory mode at a fixed strain (0.01 rad) with various angular frequencies (1–100 rad/s) as used by other studies [11,12,28]. G' of the model hydrogels consisted of SF of strain A was 4.09 ± 0.32 kPa ($n = 3$, $p < 0.01$) which was significantly higher than those consisted of SF of strains B and C at a fixed angular frequency (Table 1). Since strain A of SF consists of more amount of Tyr than those in strain B and C, it could be assumed that more di-tyrosine bones were formed to crosslink peptides of strain A of SF to produce SF-IPN hydrogels than those in B and C ones. Hence, the rheological properties of the HS-IPN hydrogels were influenced by the strain of SF. The results rheological properties influenced by strains of SF, presented in Table 1, were qualitatively consistent with those UV-excited fluorescent intensities presented in Table 2. According to amino acid analysis for three strains of SF in this study, the total number of amino acids were about 5500, including about 5.1% of Tyr in strain A, while about 4.6% of tyrosine in strains B and C. The Tyr contents for strains of B and C were similar to other reports [27]. Notably, the formations of varying amounts of dityrosine bonds in SF-IPN hydrogels, crosslinked by HRP/H_2O_2 would emit varying intensities of blue fluorescence when the hydrogels were irradiated by UV [29]. The intensity of emitted blue fluorescence of SF-IPN hydrogels for strain A was significantly higher ($n = 4$, $p < 0.01$) than those of strains B and C (Table 2), which was consistent with that Tyr contents in strain A (e.g., 277 ± 11, $n = 4$) is significantly higher than those of strain B and C (e.g., 255 ± 2, $n = 4$ for B), respectively. Therefore, the influences of strains of SF on the rheological properties of HS-IPN hydrogels possibly resulted from the tyrosine contents in each strain of SF.

Table 1. The viscoelastic parameters of the model hydrogels fabricated from different strains of SF measured at 0.01 rad and 10 rad/s ($n = 3$). The parameters of strain A such as G' and |G*|, complex shear modulus, were significantly higher than strain B and C.

Materials	Viscoelastic Properties (0.01 Rad, 10 rad/s)			
	G' (kPa)	G" (kPa)	|G*| (kPa)	δ (°)
Strain A	4.09 ± 0.32 **	0.59 ± 0.16	4.13 ± 0.34 **	8.14 ± 1.66
Strain B	3.24 ± 0.16	0.50 ± 0.05	3.27 ± 0.16	8.69 ± 0.5
Strain C	3.40 ± 0.19	0.43 ± 0.07	3.43 ± 0.18	7.28 ± 1.19

** $p < 0.01$ or better.

Table 2. The intensity of blue fluorescence of SF hydrogels excited by UV irradiation, in terms of OD values, for different strains of SF at 1.62 mM H_2O_2 [29]. ($n = 4$).

The Intensity of Blue Fluorescence of SF Hydrogels Consisted of Vary Strains of SF with Their Tyrosine Contents		
Sample	OD Value	Tyrosine Content
Strain A	0.80 ± 0.01 **	277 ± 11 **
Strain B	0.67 ± 0.01	255 ± 2
Strain C	0.67 ± 0.02	255 ± 1

** $p < 0.01$.

Influence of the Weight Ratios of SF to HA in HS-IPN Hydrogels on Rheological Properties of Hydrogels

Other than the SF strains, the weight ratios of SF to HA in producing the HS-IPN hydrogels might also influence the viscoelastic properties of the hydrogels. For examining this factor, at a fixed HA content (4%), varying SF concentrations (in wt.%) in producing HS1-IPN to HS7-IPN hydrogels (or the model gels) were carried out at the aforementioned oscillatory conditions (Figure 4A). G' values increased with increasing the concentrations of SF in the HS-IPN hydrogels. Therefore, the model hydrogels had the highest G' and G" values ($p < 0.001$, $n = 3$) among the produced hydrogels herein tested at varying angular frequencies. For example, the G' values of the model hydrogels (e.g., 4.09 kPa at 10 rad/s) were about 2.6 times higher than those of HS1-IPN gels.

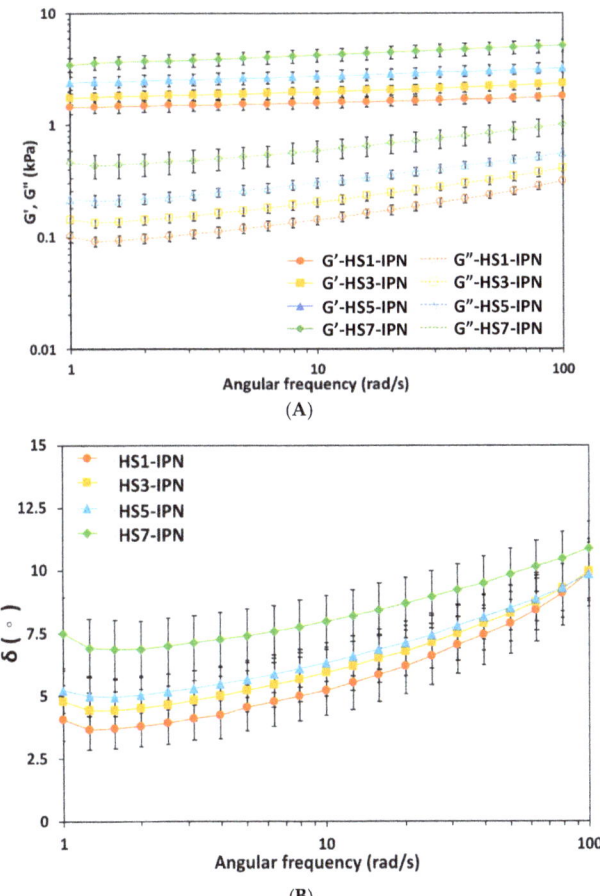

Figure 4. (**A**) The rheological parameters (e.g., G' and G") for all tested HS-IPN hydrogels increased with increasing the weight ratios of SF/HA in producing the hydrogels versus the varying angular frequencies, (**B**) increasing weight ratios of SF/HA in the HS-IPN hydrogels increased the δ values from about 4.2 (e.g., HS1-IPN) hydrogels to 8.0° (e.g., HS7-IPN) hydrogels vs. with varying angular frequencies, respectively (Data are mean ± SD, $n = 3$).

The phase angles, δ values, for all HS-IPN hydrogels were shown (Figure 4B) which increased from around 4.2° to 8.0° at 10 rad/s, respectively, ($n = 3$). The results of δ values indicated that the model hydrogels were less viscoelastic solid than others (e.g., HS1-IPN), respectively. In comparison, the rheological properties for HA crosslinked hydrogels, produced by the same protocols as those for HS-IPN hydrogels, were carried out and had low G′ (e.g., 0.24 ± 0.092 kPa, n = 3) with phase angle of 21.4°. The results indicated that viscoelastic properties for HA crosslinked hydrogels were high fluidity with a very low elastic modulus. Hence, the results of rheological properties for HA crosslinked hydrogels were not suitable for TE of NP. Notably, the G′ and δ values at 10 rad/s for the HS7-IPN hydrogels produced herein were similar to those reported for native NP (5.0~10.3 kPa and 2.5°~35°, respectively) [7,18,28], and therefore, the hydrogels were selected as the model hydrogels for further this investigation.

The G′ for IPN hydrogels produced herein were similar to those of laminin-111-PEG hydrogels reported for the matrix of NP [24]. Interestingly, the rheological modulus for G′ or |G*|of HS1-IPN hydrogels were fitted the requirements of hydrogels for cardiac repairs. According to the tyrosine contents of varying strains of SF were different (Table 2), and the strains of SF influenced the viscoelastic modules of HS-IPN hydrogels (Table 1). The contents of tyrosine of SF were one of an important factor on determining those modules of HS-IPN hydrogels although the phase angles of the hydrogels might not be the case as those modules (Table 1). Moreover, the model hydrogels contained more concentrations of SF and amounts of di-tyrosine bonds in qualitative in the hydrogels than other HS-IPN hydrogels that resulted in increasing their viscoelastic modulus (Figure 4A and Table 1). However, the amounts of di-tyrosine bonds in each HS-IPN hydrogel were not able to be determined quantitatively. Although, the bonds could be semi-quantitatively evaluated using the intensity of UV-excited blue fluorescence [29].

Although using NHS/EDC to crosslink HA/SF to facilely prepare HA/SF hydrogels was recently reported [23], it was only one step of our processes to prepare hydrogels. However, they did not perform the rheological study for the aforementioned hydrogels [23]. Recently, using sonication and UV photo-polymerization to prepare SF/methacrylated HA or to produce SF-based IPN hydrogels has been reported by Xiao et al. [33], respectively. However, the rheological properties of the hydrogels were not determined. Interestingly, using HRP/H_2O_2 to crosslink tyramine-substitute HA and 2% of SF to produce hydrogels with varying HA contents has been reported by Raia et al. [24]. HA and SF-IPN hydrogels produced in their study are configured by di-tyramine bonds in HA crosslinked network, di-tyrosine bonds in SF crosslinked network and tyramine-tyrosine bonds in HA/SF-IPN, which bonding structures of their hydrogels were distinct from those in HS-IPN hydrogels produced in this study. Hence, the rheological properties of our hydrogels were different from theirs [24]. Moreover, the tyramine-substitute HA needed to be complexly and chemically synthesized for the research which was not a commercially available biomaterial, while the biomaterials were generally commercially available.

3.4. Confined Compressive Modules of the Model Hydrogels

The confined compressive stress of the model hydrogels was conducted on the hydrogels under 5% strain with value of 0.109 ± 0.011 MPa (n = 3) which was slightly lower than the for human NP (e.g., around 0.5–1.5 MPa) [7]. Notably, the compressive modulus for the model hydrogels was 2.29 ± 0.05 MPa (n = 3) was similar to that of human NP [7]. Although using NHS/EDC to crosslink HA/SF to facilely prepare HA/SF hydrogels was recently reported by Yang et al. [23] that was generally simple to produce the hydrogels than those produced by the procedures for this study, the confined compressive stress of their products was about 12 kPa at 30% of strain, which was about 25 times less than the stress of our model hydrogels (e.g., 314 ± 2.8 kPa, at 15% strain, n = 3). Hence, the compressive stress for their hydrogels would not fit the need of human NP.

Recently, Xiao et. al. [34] reported that using sonication and further UV photo-polymerization to prepare SF-IPN and methacrylated HA to produce network SF/HA hydrogels, which had a stiff but brittle SF structure while SF-IPN hydrogels produced herein (Scheme 1) were non-brittle. However, the confined compression modules of their hydrogels were not examined [23,31] as did in this study.

3.5. Cytotoxicity Examinations for the Model Hydrogels

In-vitro cytotoxicity for the model hydrogels were performed according to the requirements of ISO10993-5. Briefly, after L929 cells were cultured with vary concentrations of extraction mediums, for 24 h, and the MTT assay to the cells were carried out to examine the cell viabilities [26]. According to the MTT assay shown in Figure 5, the L929 viability of the group with a dilution ratio of 0.5 exceeded that of the group of extraction medium only which met the requirements of ISO10993-5, revealing that the model hydrogel was biocompatible and suitable for use in TE. The results of the biocompatibility of the hydrogels were similar to other SF-based biomaterials [8,31].

3.6. Differentiations of hBMSC to NP Cells in the Model Hydrogels

To evaluate the differentiation of hBMSCs to NP cells, they were induced by TGF-β3 in the model hydrogels. The morphology of hBMSC, and the accumulations of NP-related ECM deposits in the hydrogels were examined using vimentin stain and immuno-histochemical (IHC) analysis, such as glycosaminoglycan (GAG) and collagen type II stain, respectively. The cell cultivated in cultural wells was also stained as a control group. Figure 6A exhibited that the deposition of glycosaminoglycan (GAG, in blue), one of the main ECMs in the NP, in hBMSC-laden model hydrogels was much more than that in the control [34]. Collagen type II, an important component of ECM in NP tissue, forms a fibrillar network that traps proteoglycan and resists swelling [11,12,20]. The deposition of GAG (in blue) and accumulation of collagen type II (in brown) in the model hydrogels containing differentiated hBMSC increased with increasing the cultural period (Figure 6A,B, respectively). The depositions of the GAG and collagen type II were broad distribution in the model hydrogels, revealing that the hydrogels herein were suitable to the differentiations of hBMSCs to NP cells. Enhancing expressions in GAG and collagen type II of the differentiations of hBMSCs to NP cells in the model hydrogels compared to those expressions in the control group revealed that they promoted the differentiations of hBMSC (Figure 6A,B, respectively). According to those results, the mechanical properties of the model hydrogels compared with hardness matrix (i.e., cultural wells) were suitable to the differentiations of hBMSCs to NP cells.

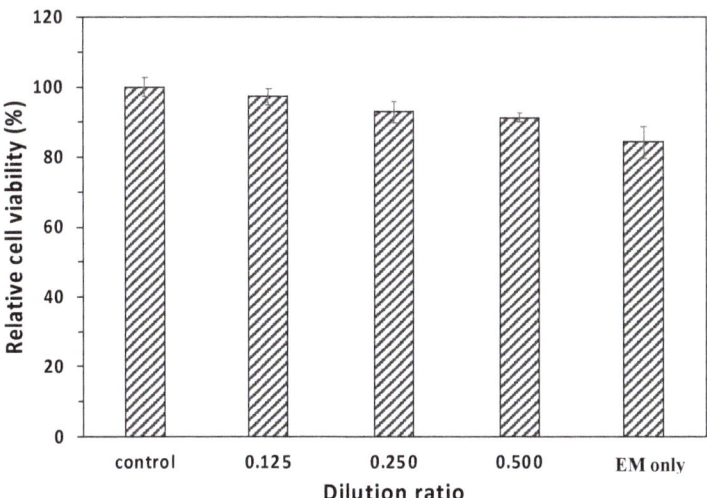

Figure 5. L929 Cell viability incubated by the extraction mediums (EM), taken from supernatants which was incubated with the model hydrogels for 24 h, at different dilute ratios according to the guidelines of ISO 10993-5. The relative L929 viability (%) of the group with a dilution ratio of 0.5 exceeded that of the group of EM only. The results of MTT assay for L929 Cell viability showed that the model hydrogels were biocompatible biomaterials (data are mean ±SD, $n = 3$).

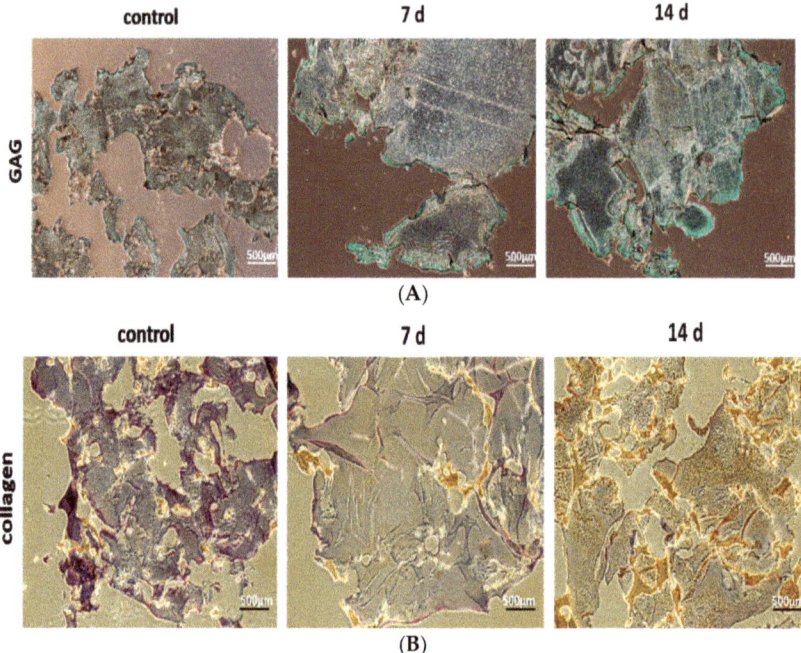

Figure 6. Immuno-histochemical (IHC) analysis of hBMSCs in the model hydrogels after hBMSC were induced by TGF-β3 to differentiate them to NP cells for 7 d and 14 d, respectively. Deposition of (**A**) GAG (in blue) and, (**B**) Collagen type II (in brown) in the model hydrogels were shown which were significantly higher than those of cultivated in cultural wells (e.g., control group).

Aggrecan (AGN), collagen type II (Col II) and collagen type I (Col II) were selected as test genes for examining the differentiations of hBMSCs in the model hydrogels [11,12,21]. The gene expressions of chondrogenesis of hBMSCs, cultured in the model hydrogels, revealed significant up-regulation of both AGN and Col II but significant down-regulation of Col I, compared to those of hBMSCs cultivated in cultural wells (e.g., the control group) for 7 d (Figure 7A–C). The results gene expressions of differentiations of hBMSCs to NP cells presented herein were similar to those of other studies [11–13] although the compositions of their hydrogels were different from this study. AGN and Col II gene expressions increased with the 7 d of differentiations of hBMSCs to NP cells in the model hydrogels that were consistent with the increasing depositions of GAG and collagen type II of IHC stains (Figure 6A,B, respectively). However, the gene expressions of the differentiations of hBMSCs to NP cells in the model hydrogels (Figure 7A–C, respectively) were not further increased in the period of 7 d to 14 d. The possible mechanisms such like reducing the activity of CD44 of HA for the differences of results the differentiations of hBMSCs to NP cells in Figures 6 and 7 at 14 d need to be further studied. Although adjust mechanical properties of HA-based or methacrylated HA-based containing SF hydrogels using varying techniques for other TE studies have been reported by others [11,23,24]. This study reported an examination of the differentiations of hBMSCs to NP cells in HS-IPN hydrogels. Nevertheless, the model hydrogels promoted the differentiations of hBMSCs to NP cells for tissue engineering of NP.

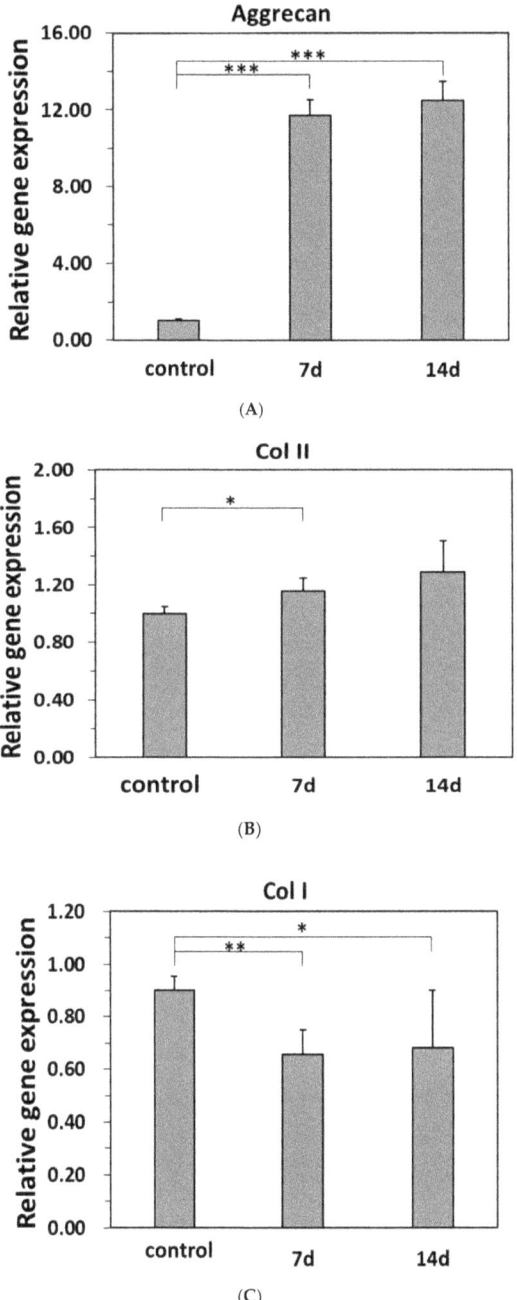

Figure 7. Real-time PCR analysis for the relative quantities of NP-specific gene expressions of inducing differentiation of hBMSC to NP cells on the model hydrogels for 7 d and 14 d. Gene expressions of AGN (**A**), COL II (**B**) and COL I (**C**) for the control group (e.g., ctrl) were the data for 7 d incubated at cultural wells. (* $p < 0.05$, ** $p < 0.01$, *** $p < 0.001$; data presented are mean ±SD, $n = 3$).

4. Conclusions

The roles of SF on characteristics of HS-IPN hydrogels were examined. The schematic procedures for preparing HS-IPN hydrogels were presented in Scheme 1. The pore size of the model hydrogels was 38.96 ± 5.05 μm between those of H and S hydrogels (Figure 2). The swelling ratios for HS-IPN hydrogels (e.g., 185.9 ± 24.4%, $n = 3$) were similar to those for H/PEI hydrogels (Figure 3) which were much less than those for HA hydrogels. The hydrogels composed of SF of strain A showed higher G' and G" values than those composed of SF of strain B and C (Tables 1 and 2, respectively). Moreover, the model hydrogels consisted of the highest weight ratios of SF to HA showed significantly higher G', G" and |G*|values than other hydrogels consisted of those of low weight ratios. (Figure 4A). The model hydrogel was biocompatible for TE applications (Figure 5). Moreover, the model hydrogels significantly promoted the differentiations of hBMSC to NP cells with increasing the expressions of GAG and collagen type II for 7 d and 14 d (Figure 6). Moreover, the induced hBMSC in the hydrogels increased the gene expressions of AGN and COL II while decreased those of COL I for 7 d of cultivations (Figure 7). Hence, the model hydrogels developed herein were suitable to tissue engineering for NP regeneration.

Author Contributions: Conceptualization, T.-W.C., W.-P.C.; methodology, T.-W.C., P.-W.T., T.-Y.W.; validation, P.-W.T., T.-Y.W.; investigation, T.-Y.W., H.-Y.L.; data curation, P.-W.T., H.-Y.L.; writing—original draft preparation, T.-W.C., W.-P.C.; writing—review and editing, T.-W.C., W.-P.C. All authors have read and agreed to the published version of the manuscript.

Funding: This research was funded by Ministry of Science and Technology of the ROC, Taiwan, for financially supporting this research under Contracts Nos. MOST 104-2221-E010-004-MY3 and 106-2221-E-027-029-MY3.

Acknowledgments: The authors would like to thank the Miaoli District Agricultural Research and Extension Station, Council of Agriculture, Executive Yuan, Taiwan for providing silk fibroin.

Conflicts of Interest: The authors declare no conflict of interest.

Nomenclature

Nomenclature	Full name
HA	Hyaluronic acid
SF	Silk fibroin
H gel	HA hydrogel
S gel	SF hydrogel
HS1-IPN hydrogel	HA/SF hydrogel with weight ratios of HA to SF was 5:1
HS3-IPN hydrogel	HA/SF hydrogel with weight ratios of HA to SF was 5:3
HS5-IPN hydrogel	HA/SF hydrogel with weight ratios of HA to SF was 5:5
HS7-IPN hydrogel (the model hydrogel)	HA/SF hydrogel with weight ratios of HA to SF was 5:7

References

1. Higuchi, A.; Ling, Q.D.; Chang, Y.; Hsu, S.T.; Umezawa, A. Physical cues of biomaterials guide stem cell differentiation fate. *Chem. Rev.* **2013**, *113*, 3297–3328. [CrossRef] [PubMed]
2. Zhang, J.; Ma, X.; Fan, D.; Zhu, C.; Deng, J.; Hui, J.; Ma, P. Synthesis and characterization of hyaluronic acid/human-like collagen hydrogels. *Mater. Sci. Eng. C* **2014**, *43*, 547–554. [CrossRef] [PubMed]
3. Wu, J.P.J.; Cheng, B.; Roffler, S.R.; Lundy, D.J.; Yen, C.Y.T.; Chen, P.; Lai, J.J.; Pun, S.H.; Stayton, P.S.; Hsieh, P.C.H. Reloadable multidrug capturing delivery system for targeted ischemic disease treatment. *Sci. Transl. Med.* **2016**, *8*, 365ra160. [CrossRef] [PubMed]
4. Lin, C.C.; Metters, A.T. Hydrogels in controlled release formulations: Network design and mathematical modeling. *Adv. Drug Deliv. Rev.* **2006**, *58*, 1379–1408. [CrossRef]
5. Ahrens, M.; Tsantrizos, A.; Donkersloot, P.; Martens, F.; Lauweryns, P.; Le Huec, J.C.; Moszko, S.; Fekete, Z.; Sherman, J.; Yuan, H.A.; et al. Nucleus replacement with the DASCOR disc arthroplasty device: Interim two-year efficacy and safety results from two prospective, non-randomized multicenter European studies. *Spine* **2009**, *34*, 1376–1384. [CrossRef] [PubMed]

6. Joshi, A.; Fussell, G.; Thomas, J.; Hsuan, A.; Lowman, A.; Karduna, A.; Vresilovic, E.; Marcolongo, M. Functional compressive mechanics of a PVA/PVP nucleus pulposus replacement. *Biomaterials* **2006**, *27*, 176–184. [CrossRef]
7. Lanza, R.; Langer, R.; Vancati, J. *Principles of Tissue Engineering*, 3rd ed.; Academic Press: Burlington, VT, USA, 2007.
8. Tsaryk, R.; Gloria, A.; Russo, T.; Anspach, L.; Santis, R.D.; Ghanaati, S.; Unger, R.E.; Ambrosio, L.; Kirkpatrick, C.J. Collagen-low molecular weight hyaluronic acid semi-interpenetrating network loaded with gelatin microspheres for cell and growth factor delivery for nucleus pulposus regeneration. *Acta Biomater.* **2015**, *20*, 10–21. [CrossRef]
9. Talebin, S.; Mehrali, M.; Taebnia, N.; Pennisi, C.P.; Kadumudi, F.B.; Foroughi, M.; Hasany, M.; Nikkhah, M.; Akbari, M.; Orive, G.; et al. Self-healing hydrogels: The next paradigm in tissue engineering. *Adv. Sci.* **2019**, *6*, 1801664. [CrossRef]
10. Yang, M.C.; Huang, Y.Y.; Wang, S.S.; Chou, N.K.; Chi, N.H.; Shieh, M.J.; Chang, Y.L.; Chung, T.W. The cardiomyogenic differentiation of rat mesenchymal stem cells on silk fibroin-polysaccharide cardiac patches in vitro. *Biomaterials* **2009**, *30*, 3757–3765. [CrossRef]
11. Su, W.Y.; Chen, Y.C.; Lin, F.H. Injectable oxidized hyaluronic acid/adipic acid di-hydrazide hydrogel for nucleus pulposus regeneration. *Acta Biomater.* **2010**, *6*, 3044–3055. [CrossRef]
12. Chen, Y.C.; Su, W.Y.; Yang, S.H.; Gefen, A.; Lin, F.H. In-situ forming hydrogels composed of oxidized high molecular weight hyaluronic acid and gelatin for nucleus pulposus regeneration. *Acta Biomater.* **2013**, *9*, 5181–5193. [CrossRef] [PubMed]
13. Kim, D.H.; Martin, J.T.; Elliott, D.M.; Smith, L.J.; Mauck, R.L. Phenotypic stability matrix elaboration and functional maturation of nucleus pulposus cells encapsulated in photo-crosslinkable hyaluronic acid hydrogels. *Acta Biomater.* **2015**, *12*, 21–29. [CrossRef] [PubMed]
14. Mehrali, M.; Bagherifard, S.; Akbari, M.; Thakur, A.; Mirani, B.; Mehrali, M.; Hasany, M.; Orive, G.; Das, P.; Dolatshahi-Pirouz, A. Bending electronics with human body: A pathway toward a cybernetic future. *Adv. Sci.* **2018**, *5*, 1700931. [CrossRef] [PubMed]
15. Affas, S.; Schafer, F.-M.; Algarrahi, K.; Cristofaro, V.; Sullivan, P.; Yang, X.; Costa, K.; Sack, B.; Gharaee-Kermani, M.; Macoska, J.A.; et al. Augmentation cystoplasty of diseased porcine bladders with silk fibroin grafts. *Tissue Eng. Part A* **2019**, *25*, 855–866. [CrossRef]
16. Chi, N.H.; Yang, M.C.; Chung, T.W.; Chen, J.Y.; Chou, N.K.; Wang, S.S. Cardiac repair achieved by bone marrow mesenchymal stem cells/silk fibroin/ hyaluronic acid patches in a rat of myocardial infarction model. *Biomaterials* **2012**, *33*, 5541–5550. [CrossRef]
17. Yan, L.P.; Silva-Correia, J.; Ribeiro, V.P.; Miranda-Gonçalves, V.; Correia, C.; da Silva Morais, A.; Sousa, R.A.; Reis, R.M.; Oliveira, A.L.; Oliveira, J.M.; et al. Tumor growth suppression induced by biomimetic silk fibroin hydrogels. *Sci. Rep.* **2016**, *6*, 31037. [CrossRef]
18. Gullbrand, S.E.; Schaer, T.P.; Agarwal, P.; Bendigo, J.R.; Dodge, G.R.; Chen, W.; Elliott, D.M.; Mauck, R.L.; Malhotra, N.R.; Smith, L.J. Translational of an injectable triple-interpenetrating-network hydrogels for intervertebral disc regeneration in a goat model. *Acta Biomater.* **2017**, *60*, 201–209. [CrossRef]
19. Lin, H.A.; Gupta, M.S.; Varma, D.M.; Gilchrist, M.L.; Nicoll, S.B. Lower crosslinking density enhances functional nucleus pulposus-like matrix elaboration by human mesenchymal stem cells in carboxymethylcellulose hydrogels. *J. Biomed. Mater. Res. Part A* **2016**, *104*, 165–177. [CrossRef]
20. Sakai, D.; Mochida, J.; Iwashina, T.; Hiyama, A.; Omi, H.; Imai, M.; Nakal, T.; Ando, K.; Hotta, T. Regenerative effects of transplanting mesenchymal stem cells embedded in aterlo-collagen to the degenerated intervertebral disc. *Biomaterials* **2006**, *27*, 335–345. [CrossRef]
21. Francisco, A.T.; Mancino, R.J.; Bowle, R.D.; Brunger, J.M.; Tainter, D.M.; Chen, Y.T.; Richardson, W.R.; Guilak, F.; Setton, L.A. Injectable laminin-functionalized hydrogel for nucleus pulposus regeneration. *Biomaterials* **2013**, *34*, 7381–7388. [CrossRef]
22. Collin, E.C.; Grad, S.; Zeugolis, D.I.; Vinatier, C.S.; Clouet, J.R.; Guicheux, J.J.; Weiss, P.; Alini, M.; Pandit, A.S. An injectable vehicle for nucleus pulposus cell-based therapy. *Biomaterials* **2011**, *32*, 2862–2870. [CrossRef]
23. Yang, S.Q.; Wang, Q.S.; Tariq, Z.; You, R.C.; Li, X.F.; Li, M.Z.; Zhang, Q. Facile preparation of bioactive silk fibroin/hyaluronic acid hydrogels. *Int. J. Biol. Macromol.* **2018**, *118*, 775–782.
24. Raia, N.R.; Partlow, B.P.; McGill, M.H.; Kimmerling, E.P.; Ghezzi, C.E.; Kaplan, D.L. Enzymatically crosslinked silk-hyaluronic acid hydrogels. *Biomaterials* **2017**, *131*, 58–67. [CrossRef] [PubMed]

25. Kenne, L.; Gohil, S.; Nilsson, E.M.; Karlsson, A.; Ericsson, D.; Helander Kenne, A.; Nord, L.I. Modification and cross-linking parameters in hyaluronic acid hydrogels-definitions and analytical methods. *Carbohydr. Polym.* **2013**, *91*, 410–418. [CrossRef] [PubMed]
26. Tuin, A.; Zandstra, J.; Kluijtmans, S.G.; Bouwstra, J.B.; Harmsen, M.C.; Van Luyn, M.J. Hyaluronic acid-recombinant gelatin gels as a scaffold for soft tissue regeneration. *Eur. Cell Mater.* **2012**, *24*, 320–330. [CrossRef]
27. Su, D.; Yao, M.; Liu, J.; Zhong, Y.; Chen, X.; Shao, Z. Enhancing mechanical properties of silk fibroin hydrogel through restricting the growth of β-sheet domains. *ACS Appl. Mater. Interfaces* **2017**, *9*, 17489–17498. [CrossRef]
28. Iatridis, J.C.; Weidenbaum, M.; Setton, L.A.; Mow, V.C. Is the nucleus pulposus a solid or a fluid? mechanical behaviors of the nucleus pulposus of the human intervertebral disc. *Spine* **1996**, *10*, 1174–1184. [CrossRef]
29. Lee, P.C.; Zan, B.S.; Chen, L.T.; Chung, T.W. Multi- functional PLGA-based Nanoparticles as a Controlled-Release Drug Delivery System for Antioxidant and Therapy. *Int. J. Nanomed.* **2019**, *14*, 1533–1549. [CrossRef]
30. Choi, S.C.; Yoo, M.A.; Lee, S.Y.; Lee, H.J.; Son, D.H.; Jung, J.; Noh, I.; Kim, C.W. Modulation of biomechanical properties of hyaluronic acid hydrogels by crosslinking agents. *J. Biomed. Mater. Res. A* **2015**, *103*, 3072–3080. [CrossRef] [PubMed]
31. Pritchard, E.M.; Hu, X.; Finley, V.; Kuo, C.K.; Kaplan, D.L. Effect of silk protein processing on drug delivery from silk films. *Macromol. Biosci.* **2013**, *13*, 311–320. [CrossRef]
32. Partlow, B.P.; Hanna, C.W.; Rnjas-Kovacina, J.; Moreau, J.E.; Applegate, M.B.; Burke, K.A.; Marelli, B.; Mitropoulos, A.N.; Omenetto, F.G.; Kaplan, D.L. Highly tunable elastomeric silk biomaterials. *Adv. Funct. Mater.* **2014**, *24*, 4615–4624. [CrossRef] [PubMed]
33. Xiao, W.Q.; Qu, X.H.; Li, J.L.; Chen, L.; Tan, Y.F.; Li, J.; Li, B.; Liao, X.L. Synthesis and characterization of cell-laden double-network hydrogels based on silk fibroin and methacrylated hyaluronic acid. *Eur. J. Polym.* **2019**, *118*, 382–392. [CrossRef]
34. Bian, L.; Zhai, D.Y.; Tous, E.; Rai, R.; Mauck, R.L.; Burdick, J.A. Enhanced MSC chondrogenesis following delivery of TGF-"β" 3 from alginate microspheres within hyaluronic acid hydrogels in vitro and in vivo. *Biomaterials* **2011**, *32*, 6425–6434. [CrossRef] [PubMed]

© 2020 by the authors. Licensee MDPI, Basel, Switzerland. This article is an open access article distributed under the terms and conditions of the Creative Commons Attribution (CC BY) license (http://creativecommons.org/licenses/by/4.0/).

Article

Biocompatibility Investigation of Hybrid Organometallic Polymers for Sub-Micron 3D Printing via Laser Two-Photon Polymerisation

Evaldas Balčiūnas [1], Nadežda Dreižė [1], Monika Grubliauskaitė [1], Silvija Urnikytė [1], Egidijus Šimoliūnas [1], Virginija Bukelskienė [1], Mindaugas Valius [1], Sara J. Baldock [2,3], John G. Hardy [2,3,*] and Daiva Baltriukienė [1,*]

[1] Institute of Biochemistry, Life Sciences Centre, Vilnius University, 10257 Vilnius, Lithuania; ev.balciunas@gmail.com (E.B.); nadezda.dreize@gmail.com (N.D.); monika.grub@gmail.com (M.G.); urniksi@gmail.com (S.U.); egidijus.simoliunas@gmail.com (E.Š.); virginija.bukelskiene@bchi.vu.lt (V.B.); mindaugas.valius@bchi.vu.lt (M.V.)
[2] Department of Chemistry, Lancaster University, Lancaster LA1 4YB, UK; s.baldock@lancaster.ac.uk
[3] Materials Science Institute, Lancaster University, Lancaster LA1 4YB, UK
* Correspondence: j.g.hardy@lancaster.ac.uk (J.G.H.); daiva.baltriukiene@bchi.vu.lt (D.B.); Tel.: +370-5223-4379 (D.B.)

Received: 29 October 2019; Accepted: 24 November 2019; Published: 27 November 2019

Abstract: Hybrid organometallic polymers are a class of functional materials which can be used to produce structures with sub-micron features via laser two-photon polymerisation. Previous studies demonstrated the relative biocompatibility of Al and Zr containing hybrid organometallic polymers in vitro. However, a deeper understanding of their effects on intracellular processes is needed if a tissue engineering strategy based on these materials is to be envisioned. Herein, primary rat myogenic cells were cultured on spin-coated Al and Zr containing polymer surfaces to investigate how each material affects the viability, adhesion strength, adhesion-associated protein expression, rate of cellular metabolism and collagen secretion. We found that the investigated surfaces supported cellular growth to full confluency. A subsequent MTT assay showed that glass and Zr surfaces led to higher rates of metabolism than did the Al surfaces. A viability assay revealed that all surfaces supported comparable levels of cell viability. Cellular adhesion strength assessment showed an insignificantly stronger relative adhesion after 4 h of culture than after 24 h. The largest amount of collagen was secreted by cells grown on the Al-containing surface. In conclusion, the materials were found to be biocompatible in vitro and have potential for bioengineering applications.

Keywords: bioactive surfaces; biomaterials; hybrid organometallic polymers; laser two-photon polymerisation; tissue engineering

1. Introduction

The concept of growing human replacement parts in the lab has been around for several decades [1,2]. Researchers have used different approaches for the engineering of artificial tissues—from allogeneic or xenogeneic tissue decellularisation [3] to approaches based on additive manufacturing [4], the latter of which offers a route to the generation of an optimal scaffold for a specific tissue type and patient which has significant potential for economic/health/societal impacts.

Laser two-photon polymerisation (LTPP) is a 3D fabrication technique capable of producing materials with fine details in their structures at sub-micron resolutions [5,6]. Various materials can be used to make 3D structures using this technique including derivatives of natural polymers (e.g., hyaluronic acid [7], gelatin [8]) and synthetic ones (e.g., derivative of polyethylene glycol [9] and SU-8 [10]).

The LTPP technique is highly versatile and allows several materials to be used in the same sample [11,12]. Compared to other fabrication techniques like stereo-lithography [13], fused deposition modelling [14,15] and selective laser sintering [16], LTPP is the only technique that allows for resolutions below the diffraction limit of light to be fabricated [17]. In addition, LTPP allows for fine-tuning of structural motifs as opposed to randomised porous structures obtained by other means, like template-casting [18] or particulate-leaching [19]. A group of interesting materials for LTPP are hybrid organometallic polymers [20]. To date, materials based on Al [21], Ge [22], Ti [23], V [24] and Zr [25] have been shown to be structurable using LTPP systems. An in vitro biocompatibility screening study showed that Al and Ti hybrids supported a comparable number of cells to glass, while the Zr-based hybrid exceeded the biocompatibility of all the other surfaces [21]. An in vivo study of our Zr-based hybrids showed them to be relatively biocompatible when implanted in rabbit muscle and that they did not cause inflammation or foreign body reaction as demonstrated by hematoxylin and eosin staining [26].

Cell-extracellular matrix interactions are among the most important processes to attenuate in attempting to recreate structurally and functionally viable tissue constructs analogous to natural tissues (where the resident cells adhere to the ECM and take part in remodelling it over time). For many applications, it is important to have materials that support a native-like cellular response and integration. To understand the influence of the metals contained in the organometallic hybrids on cellular behaviour, we have investigated the process of adhesion and adhesion-associated kinase expression, as well as collagen secretion of primary rat myogenic cells grown on these surfaces. We believe that this work will prove useful for tissue engineering researchers focusing on artificial scaffold-based tissue remodelling techniques in that the materials investigated in this work are highly biocompatible, precisely structurable at sub-micron resolutions and simple to prepare.

2. Materials and Methods

2.1. Material Synthesis

The hybrid organometallic polymers were synthesised according to published protocols [21,25] Briefly, the Al-based material was prepared by dissolution of aluminium isopropoxide (AIP, ≥98%, Merck, Kenilworth, NJ, USA) in toluene (ACS, ISO, Reag. Ph Eur, Merck, Kenilworth, NJ, USA). In parallel, 3-(trimethoxysilyl)propyl methacrylate (MAPTMS, 98%, Sigma-Aldrich, St. Louis, MO, USA) was hydrolysed using HCl (0.1 M, Applichem, Darmstadt, Germany). Methacrylic acid (MAA, 99%, Sigma-Aldrich, USA) was then added to the solution of aluminium isopropoxide in toluene at a 1:1 molar ratio and subsequently, hydrolysed MAPTMS was added to the mixture at a 1:1:4 AIP:MAA:MAPTMS molar ratio. Finally, 1% of photoinitiator (4,4'-bis(diethylamino)benzophenone, Sigma-Aldrich, St. Louis, MO, USA) was added to the weight of AIP, MAA and 3-(trihydroxysilyl)propyl methacrylate (the product of MAPTMS hydrolysis) and stirred, while shielding from ambient light to prevent undesired crosslinking.

The Zr-based material was prepared in an analogous manner, with molar ratios of zirconium (IV) propoxide, MAA and MAPTHS being 1:1:4 with 1% of photoinitiator by weight (excluding solvents).

2.2. 2D Sample Preparation

To secure polymer bonding to glass, MAPTMS-treated glass slides were prepared according to a protocol adopted from Kapyla et al. [27]. Briefly, circular 12 mm diameter borosilicate glass slides (Thermo Fisher Scientific, Waltham, MA, USA) were washed in ethanol, then immersed in a mixture of MAPTMS/ethanol/acetic acid/water overnight and finally washed in ethanol in an ultrasonic bath (EMAG, Mörfelden-Walldorf, Germany).

The hybrid materials were then spin-coated @ 3000 rpm for 30 s per sample and left at room temperature in the dark overnight for the solvents to evaporate. The samples were then polymerised using a UV lamp (UV-C, G15W T8, Sylvania, London, UK) for at least 5 h at around 20 cm distance, corresponding to about 1.8 mW/cm^2. The samples were sterilised under UV for at least one hour on each side, subsequently washed in sterile PBS to remove residual initiator and low molecular weight components. They were then immersed in sterile growth medium prior to cell culture.

2.3. 3D Sample Fabrication via Laser Two-Photon Polymerisation

For laser two-photon polymerisation, a droplet of a hybrid material was placed on a glass slide and then left in a fume hood to evaporate overnight while covered from ambient light. After 24 h such samples were used for structure fabrication on a commercially available Nanoscribe® Photonic Professional GT system (Nanoscribe GmbH, Eggenstein-Leopoldshafen, Germany) that is based at the Department of Chemistry at Lancaster University. A 63X 1.4 NA oil immersion lens was used to focus the laser beam into the polymer. Stereolithography (STL) flies were downloaded from the internet [28] and sliced using proprietary Nanoscribe software (DeScribe 2.3.3). Hatching and slicing distances in X, Y and Z coordinates were set to 100 nm. The laser scanning speed was set to 10 mm/s and the laser power was in the range of 20% to 100% (corresponding to 10–50 mW). After polymerization, the Al based hybrid material was developed in toluene for at least 15 min, followed by air-drying. The structures were then sputter-coated at a thickness of 10 nm using a Q150 RS coater (Quorum Technologies, Lewes, UK) and subsequently observed using a SEM (JSM 7800F, JEOL, Tokyo, Japan) operating at 10–15 kV at the Department of Chemistry at Lancaster University.

2.4. Collagen Adsorption and Secretion Assays

A protocol for assessing the concentration of collagen was adapted from Tullberg-Reinert and Jundt [29] with slight modifications. The protocol used is based on in situ staining of cell monolayers using Sirius Red. This dye has strong interactions with collagen types I and III, but weak interaction with collagen type IV.

First, a solution of collagen type I (1 mg/mL in 0.1 N acetic acid, Sigma-Aldrich, St. Louis, MO, USA) was added to the sample and incubated for 1 h at 37 °C. Then, the collagen solution was gently replaced with fixating Bouin's solution (15 mL saturated aqueous picric acid (Sigma-Aldrich, St. Louis, MO, USA) with 5 mL 35% formaldehyde (Sigma-Aldrich, St. Louis, MO, USA) and 1 mL glacial acetic acid (Sigma-Aldrich, St. Louis, MO, USA), incubated for 1 h and then washed using PBS. The samples were then transferred to new wells with a solution of Sirius Red (1 mg/mL in picric acid) for 1 h under mild rocking conditions (30 rpm). Afterwards, the samples were washed with 0.01 N hydrochloric acid to remove unbound dye. The dye was then dissolved in 0.2–0.3 mL of 0.1 N sodium hydroxide by shaking at room temperature for 30 min. Non-coated samples were stained and measured in the same way and their signals were subtracted from collagen-coated sample signals prior to analysis.

The optical density of the solution was measured using a Varioskan Flash microplate reader (Thermo Fisher Scientific, Waltham, MA, USA) at 550 nm. For reference, a calibration curve was prepared by using known concentrations of collagen type I in order to link them to the absorbance of Sirius Red at 550 nm (Figure S1 in Supplementary Materials). A total of 3 independent experiments were carried out and the results are presented as averages +/- standard errors.

Collagen synthesis was assessed in an analogous way to adsorption. The cells were seeded on the test surfaces at a density of 20,000 cells/mL/sample in 24 well tissue culture plates. After 1, 7 and 14 days of incubation, the samples were fixated using Bouin's solution, then dyed using a solution of Sirius Red (Sigma-Aldrich, St. Louis, MO, USA) and subsequently washed with a solution of HCl (Sigma-Aldrich, St. Louis, MO, USA). Finally, the dye was dissolved in a solution of NaOH (Merck, Kenilworth, NJ, USA) and measured spectrophotometrically against pure NaOH, and the amount of collagen was determined from the calibration curve. 3 independent experiments were performed with 3 repetitions per material within each experiment.

2.5. Cell Isolation, Culture and Characterisation

A Wistar rat was euthanised and a small piece of skeletal muscle tissue was removed. The experiments were approved by License of Animal Research Ethics Committee (Lithuania) No. G2-39, 03/08/2016. Large blood vessels were separated, then skeletal muscle tissue was minced and incubated in a solution of EDTA-trypsin (Gibco, Thermo Fisher Scientific, Waltham, MA, USA) with collagenase and hyaluronidase (0.5% and 0.3%, respectively, Sigma-Aldrich, St. Louis, MO, USA) for 30 min at 37 °C under mild shaking conditions. The resulting cell suspension was centrifuged, mixed with growth medium and seeded to tissue culture plates. The growth medium was replaced every 3–4 days and after 5 passages the cells were cloned by serial dilution. A highly proliferative colony forming unit was selected as a cell source, multiplied in vitro and banked in liquid nitrogen for further use.

The cells were cultured in Iscove's modified Dulbecco's medium (IMDM), supplemented with 10% Foetal bovine serum (Gibco, Thermo Fisher Scientific, Waltham, MA, USA) and Penicillin-Streptomycin (100 U/mL and 100 µg/mL, Gibco, Thermo Fisher Scientific, Waltham, MA, USA). They were subcultured every 3–4 days, by detaching with EDTA-trypsin and resuspending in fresh medium. The cells were grown in an incubator (Thermo Fisher Scientific, Waltham, MA, USA) at 37 °C with 5% CO_2.

For characterization, cells were grown in 30 mm diameter Petri dishes with glass slides on the bottom, fixed with 4% paraformaldehyde for 15 min. Then the cells were washed with PBS and incubated with 0.2% Triton X-100 in PBS for 15 min to permeabilise the membranes. After blocking in 1% BSA in PBS, the cells were incubated with primary antibodies against CD34, C45 (both from Abcam, Cambridge, UK), Myf5 and c-kit (both from Thermo Fisher Scientific, Waltham, MA, USA) overnight according to manufacturer's instructions at 4 °C. The samples were then rinsed with 1% BSA in PBS and incubated with Cy3-conjugated secondary antibodies (Merck, Kenilworth, NJ, USA). Analysis was performed using an Olympus IX71 (Olympus, Tokyo, Japan) fluorescence microscope.

2.6. Analysis of Cell Viability

AO/EB staining distinguishes cells into four categories: live (bright green, intact nucleus), early apoptotic (bright green, but fragmented nucleus), late apoptotic (orange, fragmented nucleus) and necrotic (bright orange, intact nucleus). Cells were cultured on the hybrid polymer and glass surfaces for 24, 48, 72 and 96 h. Then, the growth medium with any unattached cells was collected and the monolayer of cells were treated with EDTA-trypsin and then collected as well. All of them were subsequently stained by acridine orange and ethidium bromide (Sigma-Aldrich, St. Louis, MO, USA) as described in [30]. The cells were analysed using a BD FACSCanto™ II (BD Biosciences, San Jose, CA, USA) flow cytometer, registering 10,000 events per sample and time point. Green fluorescence of acridine orange was detected using FITC channel and red fluoresce of ethidium bromide was detected using PE channel. A total of 3 independent experiments were performed with 3 repetitions per material within each experiment. The cells were split into four categories based on their fluorescence profile.

2.7. MTT Assay

MTT assay is based on cellular mitochondrial reductase activity. The absorption measured using this technique is directly proportional to the amount of active enzyme, which gives an idea on the number of viable cells and their metabolic activity.

The cells were seeded at a density of 20,000 cells/mL/sample in 24 well tissue culture plates on the hybrid material and glass surfaces and cultured for 24, 48, 72 and 96 h at 37 °C. The samples were then gently transferred to new tissue culture plates containing a solution of MTT (0.2 mg/mL in PBS) and incubated for 2 h at 37 °C. The solution was subsequently removed and the formazan crystals dissolved in ethanol (96%, Vilniaus degtinė, Vilnius, Lithuania). The absorption was measured using a Varioskan Flash microplate reader (Thermo Fisher Scientific, Waltham, MA, USA) at 570 nm with ethanol as a reference. A total of 3 independent experiments were performed with 3 repetitions per material within each experiment.

2.8. Adhesion Strength

The cells were seeded at a density of 100 000 cells/mL/sample in 24 well tissue culture plates and cultured for either 4 or 24 h at 37 °C with 5% CO_2. At those time points, samples were transferred to new tissue culture plates, and half of the plates were subjected to shaking at 500 rpm for 5 min using a tissue culture plate shaker (Thermomixer Comfort, Eppendorf, Hamburg, Germany). The reference samples (unshaken) were incubated at 37 °C. The number of cells remaining adhered to the surface was measured by replacing the growth medium with 0.1% crystal violet solution in 20% ethanol for 30 min. Then cells were washed with tap water. Before the measurement of absorption, dye was solubilised with 0,1% acetic acid solution in 50% ethanol. Optical density proportional to cell number was measured using a Varioskan Flash microplate reader (Thermo Fisher Scientific, Waltham, MA, USA) at 570 nm and comparing it to analogously dyed, unshaken cell monolayers. A total of 3 independent experiments were performed with 3 repeats per material within each experiment.

2.9. Signalling Protein Expression and Phosphorylation

The cells were seeded on the samples at a density of 100,000 cells/mL/sample and cultured for either 4 or 24 h. At these timepoints, the samples were transferred to new tissue culture plates and gently washed with PBS. The PBS was then replaced with a lysis buffer consisting of 8 M urea, 2 M thiourea and 50 mM DTT. Cells were collected from 5 samples for each timepoint by pipetting 5–10 times. The lysates were then centrifuged for 10 min at 20,000 × G at RT. The supernatants of each vial were then transferred to new vials and frozen at −20 °C until further use. A total of 3 independent experiments was carried out.

Protein concentrations were normalised by running an SDS-PAGE gel, staining with Coomassie brilliant blue, taking images using a transilluminator (UVP, Upland, CA, USA) using analysing by ImageJ software. After diluting the highest concentration samples using lysis buffer, the concentration-equalised samples were subjected to gel electrophoresis again at 200 V for 45 min using a BioRad (Hercules, CA, USA) electrophoresis apparatus. The proteins were then transferred to a PVDF membrane (Carl Roth, Karlsruhe, Germany) using a Biometra Fastblot transfer device (Biometra GmbH, Göttingen, Germany) at 25 V and 300 mA. The membranes were blocked using 1% BSA (Sigma-Aldrich, St. Louis, MO, USA) in TBS with 0.1% Tween 20 (Sigma-Aldrich, St. Louis, MO, USA).

The membranes were subsequently treated with primary antibodies against p-Akt (Ser473 and Thr308, Cell Signalling technology, Danvers, MA, USA), Akt (Molecular Probes, Eugene, OR, USA), FAK (BD Biosciences, San Jose, CA, USA) and fluorescently-labelled anti-α-tubulin (Sigma-Aldrich, St. Louis, MO, USA) overnight according to manufacturer's instructions at 4 °C.

Next day, the membranes were washed three times using wash buffer and incubated with secondary antibodies. FAK and Akt were treated with HRP-conjugated goat anti-mouse (Invitrogen, Carlsbad, CA, USA) and HRP-conjugated anti-rabbit (Invitrogen, Carlsbad, CA, USA) according to manufacturer's instructions for 1 h. Finally, the membranes were washed again three times using wash buffer. FAK and Akt were detected using by treating with ECL reagent, which upon catalysis by HRP yields chemiluminescence that was detected using a transilluminator (UVP, Upland, CA, USA).

P-Akt (Ser473) and p-Akt (Thr308) were treated with secondary goat-anti-rabbit antibodies conjugated to a fluorescent infrared dye (IRDye 800CW, LI-COR, Lincoln, NE, USA) and detected using an infrared imaging system (Odyssey, LI-COR, Lincoln, NE, USA).

Membrane image analysis was performed using ImageJ software (National Institutes of Health, USA).

2.10. Statistical Analysis

All experiments were repeated at least 3 times independently with 3 repetitions within each experiment. The results are presented as averages +/− standard deviations (n = 0–5) or errors

($n = 6+$). Statistical significance was assessed using one-way or two-way ANOVA and Tukey's HSD post-hoc tests using RStudio (RStudio Inc., Boston, MA, USA) and plotted using the ggplot2 package. Statistical significance was considered to be achieved with $p < 0.05$.

3. Results and Discussion

3.1. Laser Two-Photon Polymerisation

As mentioned in the introduction and methods, the possibility of structuring the Al-based hybrid material has already been described previously [21]. However, we wanted to further demonstrate our ability to fabricate complex 3D shapes and thus have fabricated a micro-structure reminiscent of Zerg (StarCraft, Blizzard Entertainment©, Irvine, CA, USA)—see Figure 1. High volume suspended features were reproducibly fabricated, like the claws of the Zerg hydralisk. Literature shows that by precisely controlling the 3D cell microenvironments, one can achieve complex cellular responses, like stem cell homing towards artificially designed niches [31]. Another important group of factors in designing artificial cell niches are associated to the material chemistry and properties [32]. Integration of several materials in a single structure will hopefully one day help to guide stem cell differentiation towards different lineages on the same sample.

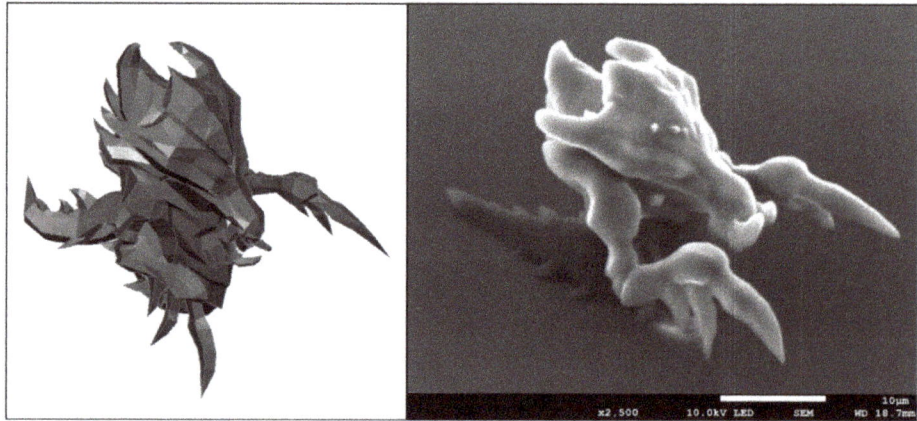

Figure 1. Right, A 3D micro-structure reminiscent of Zerg fabricated using a commercially available Nanoscribe Photonic Professional GT system. The scale bar corresponds to 10 microns. Left, A CAD model of the structure.

Multiple materials can be structured with laser two-photon polymerisation and the Al-based hybrid organometallic polymer presented in this work expands the selection of possible cellular niche materials even further. To date, a number of reports have been made of multiple-material structure fabrication using laser two-photon polymerization—including some of our own, in which PDMS, hybrid organometallic polymer based on Zr, PEG-DA and commercially-available OrmoComp have been integrated together [12]. Lamont et al. have developed a technique to exchange the material that is being used to fabricate structures on-the-go, allowing up to five different materials to be incorporated within the same structure with increased fabrication speed and potentially, multiple functionalities for applications like drug delivery, advanced optics, meta-materials and microrobotics [33]. Additionally, two or even more disparate fabrication techniques can be employed within the same structure as demonstrated with fused filament fabrication 3D-printed structure modification using laser ablation [34].

The addition of yet another 3D-printable material with new properties to the engineer's menu will allow for a wider variety of new structures and applications to be envisaged, particularly in fields that require high precision micro-fabrication, like tissue engineering scaffolds and drug delivery devices.

3.2. Protein Adsorption Assay

Extracellular matrix protein adhesion to the different surfaces may affect the subsequent adhesion of cells to the surfaces. Collagen type I was chosen as a model system for this experiment—its adsorption to simple spin-coated samples was assessed due to collagen type I abundance in the extracellular matrix. Collagen supports cellular adhesion via RGD sequences, which are specifically recognised by cellular adhesion proteins integrins. These interactions facilitate strong integration between a tissue engineered scaffold and the surrounding tissues, so it is important to have a material that supports collagen adsorption to its surface well for applications where cell adhesion is beneficial.

Collagen itself has been used as a tissue engineering building block in various studies, for example that of Ber et al., where osteoblast growth was guided on 2D surfaces of collagen [35] or work by Ramanathan et al., who have investigated 3D hybrid collagen matrixes as antibacterial dermal substitutes [36]. Collagen and its derivatives are a widely investigated group of materials for soft tissue engineering, while for hard tissue, like bone, collagen coating is a highly desirable approach that can be used in conjunction with other materials. Multiple reviews have focused on the use of collagen as a tissue engineering material either in a pure or composite form [37–39]. Consequently, a possibility of coating hybrid polymer structures with collagen seems like a good approach for improving cellular integration and therefore chosen in this study.

In our experiments, the choice of protein quantification technique was limited by residual photo-initiators within the materials which rendered them fluorescent. Thus, the signal of fluorescence-based techniques for protein visualisation would have been hindered by sample autofluorescence. To overcome this issue, a colorimetric assay was employed.

Collagen adsorption was measured on both the hybrid polymers and reference glass slides by immersion in a solution of collagen, then dying the bound collagen and measuring the optical absorption of the dye. We found that even though the largest amount of collagen was found on the Zr surface and the smallest amount on Al surface, the results in each group were highly disperse, and no statistically significant differences were determined with $p = 0.35$ between Al and Zr, $p = 0.72$ between Glass and Zr and $p = 0.81$ between Glass and Al. The results are presented in Figure 2.

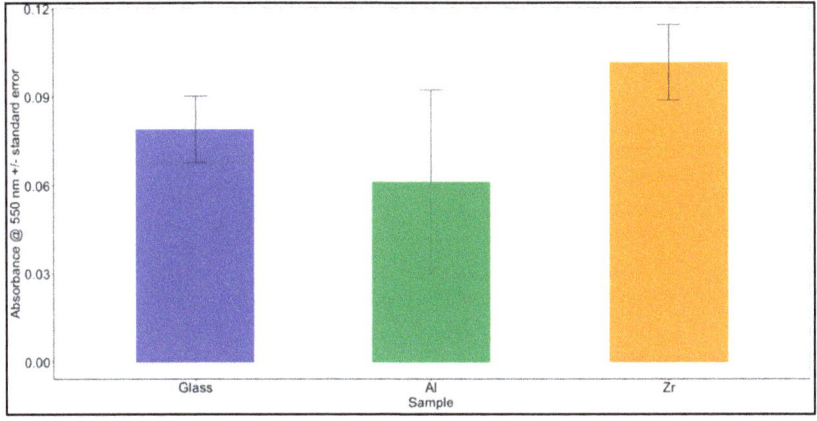

Figure 2. Collagen I adsorption to Al and Zr based hybrid organometallic polymers and glass. The results are presented as average Sirius Red absorbances + standard errors for three independent experiments with n = 3 per each experiment (a total of n = 9 per material).

A calibration curve between the total amount of collagen I and Sirius Red absorption was prepared and is presented in Supplementary Materials Figure S1. A highly linear relation was found in the range of 0–100 µg with an R^2 value of 0.994. The amounts of collagen adsorbed to the samples were calculated according to the linear model and were found to be 5.59 µg for Glass, 4.40 µg for Al and 6.73 µg for Zr on average per sample.

In our previous work [21], we investigated the contact angles of these materials and found them to be 39° for Glass, 72° for Al-based hybrid materials and 71° for Zr-based hybrid materials. The collagen adsorption data does not seem to correlate with the surface contact angles. Even though there is a significant difference between the contact angle of glass to Al or Zr surfaces, the difference in collagen I adsorption could not be distinguished due to a high deviation from the mean. A paper by Ying et al. [40] indicates that around a two-fold increase in the adsorption of collagen I to the glass surface is expected with an increase of the surface contact angle from 40° to 70°. An analogous result would be expected on other surfaces, since protein adsorption is governed by the same forces as surface contact angle—electrostatic and hydrophobic interactions.

3.3. Cell Viability

Having established the structurability of the materials as well as their collagen adsorption capacity, we investigated how their surfaces influence cellular viability. In this particular study, a primary rat muscle cell line was used to model a situation that would be as close as possible to a clinical one. Using primary muscle-derived stem cells is a safe choice in designing a possible future tissue engineering strategy due to the abundance of donor tissue sites, relatively mild donor site morbidity and high proliferative capacity as well as multipotency of the cells. A great example by Nieponice et al. demonstrates how rat muscle-derived stem cells could be used together with elastomeric scaffold-building materials in constructing vascular grafts [41]. The study successfully showed that the cells tended to home to the lumen of the vascular grafts and started to express α-actin, calponin as well as secrete collagen. Endothelial differentiation was supported via the presence of von Willebrand factor. Our isolated rat muscle-derived progenitor cells were positive for CD34, Myf5 and c-kit and negative for CD45 (Figure S2).

This experiment aimed to find the ratios between the numbers of viable, early apoptotic, necrotic and late apoptotic cells at different timepoints and on different surfaces. Flow cytometry data shows that no statistically significant differences between the numbers of cells in each category can be observed, since the majority of cells at each time point was viable (Figure 3). Additional data is presented in the Supplementary Materials, providing light microscopy images of cell culture at different time points as well as statistical analysis (Figure S3 and Figure S4 in the Supplementary Materials).

The results are in accord with our previous work on these materials with NIH/3T3 cells, which showed that the cells remain viable in culture for extended periods of time [21]. Data obtained in this experiment open the way for the future applications of both Al and Zr based hybrid materials. Long-term viability support of the cells is the first step towards engineering adequate implants for patients.

3.4. Cell Proliferation and Metabolism

As a subsequent step after viability assessment, the rate of cellular metabolism was measured using MTT assay. This test is a direct indicator of mitochondrial activity, since the dye is being digested by mitochondrial reductase to yield a colorimetric signal. This can be used as an indicator for overall cellular metabolism. It is important to know if cellular metabolism rates are different among the various surfaces, since significant deviations in cellular metabolism rates could potentially yield new properties of the cell population on that particular surface. Increased rates of metabolism could be associated with malignant formations and dysregulation of the cell cycle, while down-regulated rates of metabolism could indicate low cellular viability or transfer into a more quiescent state via differentiation or other mechanisms [42].

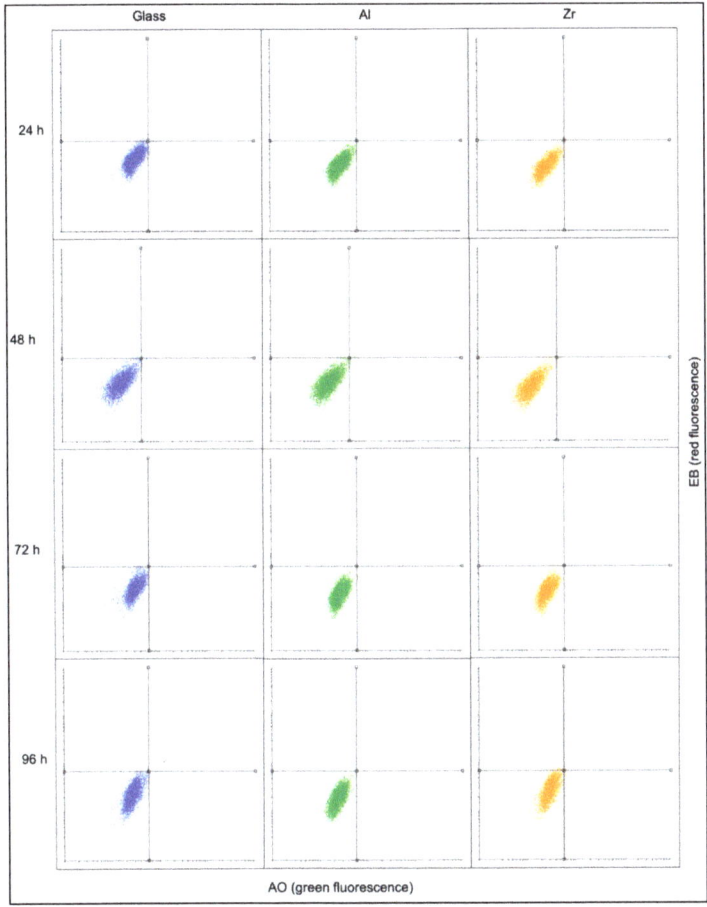

Figure 3. Flow cytometry data on cellular viability on Glass, Al and Zr surfaces after 24, 48, 72 and 96 h. The four quadrants in each case show the following: lower left—viable cells, upper left—necrotic cells, upper right—late apoptotic cells, lower right—early apoptotic cells. The lower left quadrant corresponds to 95% of all signals in all cases.

Light microscopy images show cells grown on both the hybrid polymer and glass surfaces to have formed confluent monolayers after 48 to 72 h (Supplementary Materials, Figure S3). No significant differences were observed neither in daily culture check-ups nor in the images taken of those cells. The cells had healthy spindle-shaped morphologies and stopped dividing after having reached confluency.

The assessment of total metabolic activity revealed that the metabolic rate of cells grown on glass was significantly higher than on the hybrid organometallic polymers after 96 h of culture ($p < 0.001$) and significantly higher than on the Al-based hybrid starting from 48 h ($p < 0.05$). These results are presented in Figure 4. Our previous study [21] investigated the proliferation rate of NIH/3T3 fibroblasts on analogous surfaces after 120 h and showed the Zr surface to support the highest rate, while no statistically significant differences were observed between Glass and Al.

The results obtained in this experiment show promise in using both the Al and Zr hybrid polymers in tissue engineering applications. Both materials supported cell growth, even though the total rate of metabolism was lower than on reference glass surfaces.

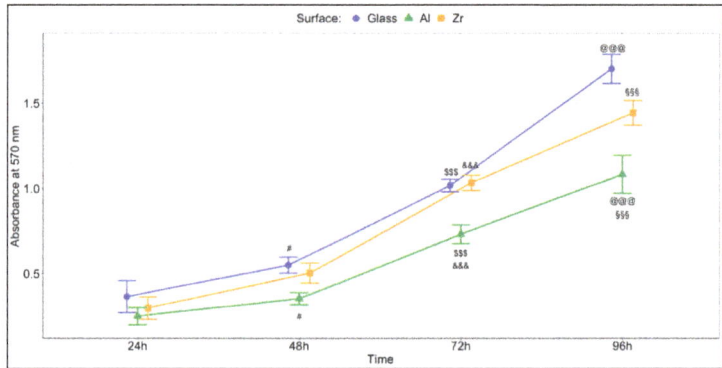

Figure 4. Cell proliferation as measured by MTT. The results are presented as average +/- standard error. A total of 9 measurements were performed per material per time point (3 independent experiments with 3 repetitions in each). #—significant difference between glass and Al with $p < 0.05$; \$\$\$—significant difference between glass and Al with $p < 0.001$; &&&—significant difference between Al and Zr with $p < 0.001$; @@@—significant difference between glass and Al with $p < 0.001$; §§§—significant difference between Al and Zr with $p < 0.001$.

3.5. Cell Adhesion Strength

Cellular viability, proliferation and adhesion are highly intertwined processes. Knowing that the viability does not change with surface and that the rate of cellular metabolism grown on the Al hybrid is lower, we sought to understand if the reduced metabolic activity was associated with weaker adhesion. The cells were seeded on the samples and then shaken vigorously at two timepoints. The number of remaining cells was counted and calculated against the number of cells on reference surfaces without shaking. The results are presented in Figure 5. There was a clear trend that showed a decrease in the number of adhered cells after 24 h, suggesting that the initial stage of attachment after 4 h was stronger. Even though there were substantial differences between the number of adhered cells after 24 h, the differences were not statistically significant. However, in spite of a lack of statistical significance, the number of cells that remained attached to the Al surface was the lowest and thus in accord with the proliferation data, that showed the lowest rate of MTT metabolism on the Al- based hybrid. Overall, the Al surface seems to be slightly less supportive of cell adhesion, though insignificantly. The reason for that could potentially be the lowest adsorption of collagen to the Al surface, as presented in Figure 2. To further investigate this difference, an experiment was devised to assess the expression of adhesion-associated proteins.

3.6. FAK/Akt Expression and Activation

When integrin-mediated cell adhesion takes place, a cascade of biochemical reactions is activated that delivers a signal about cellular attachment to the cell nucleus, which in turn activates the transcription and translation of certain proteins that are required for these processes. The cascade is activated when cell surface integrins dimerise after binding to a specific adhesion sequence in the ECM [43]. To initiate intracellular signalling integrin dimer induces conformational changes in focal adhesion kinase (FAK), which is one of the first proteins to join the newly forming focal adhesion. Subsequently, this leads to the recruitment of multiple other adhesion-associated proteins to the focal adhesion site. Upon formation of a focal adhesions, multiple pathways can be activated, one of which is the Akt pathway. This pathway is responsible for transferring a survival and proliferation signal to the nucleus via the mTOR pathway [44]. Akt can be activated by either phosphorylation on Ser473 or Thr308—this depends on the phosphorylating protein. In the case of integrin-mediated adhesion,

PI3K is being activated and thus, only Ser473 is phosphorylated [45]. Phosphorylation on the Thr308 occurs from mTORC2, which is associated with cellular metabolism and cytoskeletal reorganisation.

To assess the expression and activation of adhesion-associated FAK pathway proteins, a Western blot analysis was performed on cells grown on the investigated Glass, Al and Zr surfaces. The results are presented in Figure 6 with membrane images presented in Supplementary Materials Figure S5. Even though no statistically significant results were observed on the investigated surfaces, a tendency for reduced phosphorylation of Akt kinase can be seen in most of the cases when comparing 4 and 24 h. This is to be expected as the adhesion-associated Akt kinase's role is mostly pronounced during the initial cell-to-surface and cell-cell interactions that take place upon cell seeding.

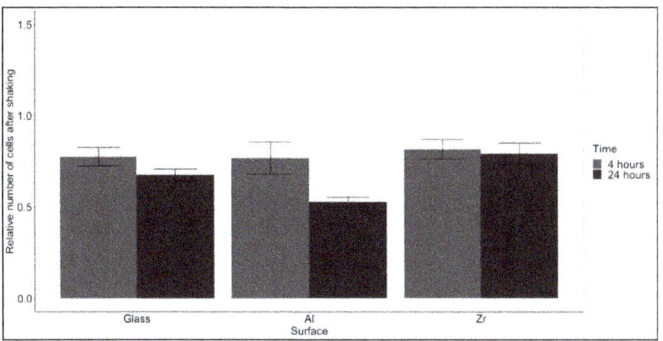

Figure 5. Evaluation of cell adhesion strength on the tested surfaces. The results are presented as a ratio between the average number of cells after shaking to that of cells before shaking after 4 and 24 h of culture + standard errors. A total of 9 measurements per material were performed (3 independent experiments with 3 repetitions each). No statistically significant differences were observed between the different time points on each surface.

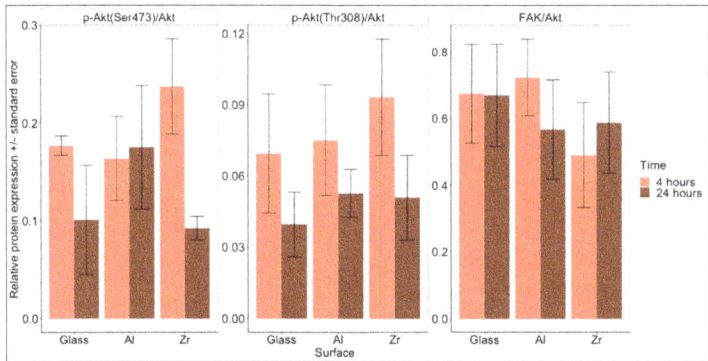

Figure 6. Western blot data, showing the expression of FAK and the expression and activation of Akt kinase after 4 and 24 h of culture on glass, Al and Zr surfaces. The amount of p-Akt (Ser 473), p-Akt (Thr308), Akt and FAK were observed. The results are presented as relative expression +/- standard error. Three independent experiments were performed. No statistically significant differences were observed.

The ratio between FAK and Akt kinases was mostly unchanged between time points and only slightly and insignificantly lower on Zr surfaces. This is again to be expected as the total amount of FAK and Akt kinases should remain unchanged.

On the glass surface, the cells tended to attach stronger during the initial 4 h—this is in accord with a higher level of Akt kinase activation in this time point. After 24 h the drop in the number of adhered cells and p-Akt (both Ser473 and Thr308) is clearly visible, though insignificant.

On the Al surface, the situation is similar in terms of p-Akt (Thr308), where the amount of phosphorylated Akt drops after 24 h in accord with a decreased number of cells. However, this was not observed in the p-Akt (Ser473) test. The level of FAK decreased as well, suggesting that the number of focal adhesions had been reduced as well.

The highest drop in terms of level of phosphorylation was observed on the Zr surface. From 4 to 24 h, the level of p-Akt (Ser473) dropped with a $p = 0.25$ and the level of p-Akt (Thr308) dropped with a $p = 0.67$. A slight, but insignificant increase was observed in the level of total FAK. Even though the changes in phosphorylation of Akt on Zr surfaces was the highest, the change in observed cell detachment was lowest in this case.

Even though some tendencies in cell attachment and adhesion-associated protein expression can be seen, these are not statistically significant.

Strong cellular adhesion is an essential prerequisite for successful tissue engineering techniques as discussed in a review by Lee et al. [46]. Whenever adhesion-associated signalling is insufficient, anchorage-dependent cells undergo anoikis, a form of programmed cell death that occurs in anchorage-dependent cells when they detach from the surrounding ECM.

3.7. Collagen Synthesis

Finally, we wanted to investigate whether the surface had any effect on the synthesis of collagen type I. Collagen is one of the main components of the extracellular matrix. Healthy tissues constantly undergo remodelling of their protein structures with the creation and digestion of new proteins by parenchymal cells, like fibroblasts. In tissue engineering, it is essential to have cells be able to model their environments by secreting ECM proteins, like collagen I.

Even though the adsorption was weakest to the Al-based hybrid and so was the cellular adhesion and proliferation, the synthesis of collagen in cells grown on the Al hybrid was the highest. The results are presented in Figure 7. The difference is already visible after one week, but it becomes significant after 2 weeks. This is to be expected as the process of extracellular matrix remodelling takes a substantial amount of time.

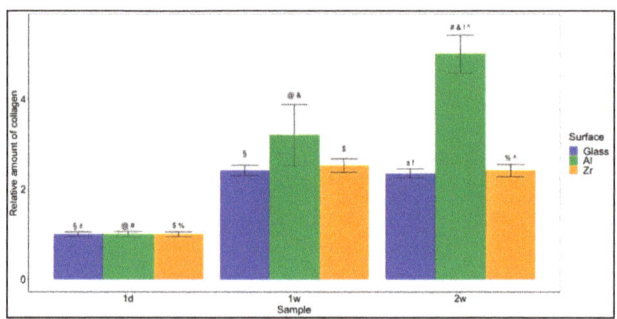

Figure 7. Collagen synthesis. Relative amount of collagen produced by cells grown on the different surfaces after 1 day, 1 week and 2 weeks. The results are presented as averages +/- standard errors. Statistically significant differences are indicated using different symbols. §—$p < 0.05$ between 1 day and 1 week glass; ±—$p < 0.05$ between 1 day and 2 week glass; @@@—$p < 0.001$ between 1 day and 1 week Al; ###—$p < 0.001$ between 1 day and 2 week Al; $$—$p < 0.01$ between 1 day and 1 week Zr; %—$p < 0.05$ between 1 day and 2 week Zr; &&&—$p < 0.001$ between 1 week and 2 week Al; !!!—$p < 0.001$ between 2 week glass and 2 week Al; ^^^—$p < 0.001$ between 2 week Al and 2 week Zr. Other statistically significant differences (those between both different materials and different timepoints) are not presented.

Other statistically significant differences were observed between Al surfaces after 1 day, 1 week and 2 weeks with $p < 0.001$ between all periods. As well as Zr surfaces after 1 day and 1 week with a $p < 0.05$.

The results show promise for tissue engineering applications, because they show a significant increase in the collagen produced in the cells grown on these surfaces. This means that the cells had not only attached to the surfaces, but they are comfortable and viable enough to start remodelling their environment via collagen synthesis.

4. Conclusions

We believe that the materials investigated in this work may be useful in constructing tissue engineered grafts of the future. This study provides a direct comparison between two materials that can be structured for laser two-photon polymerisation, assessing their support of primary rat myogenic cell growth. The differences between Zr and Al surfaces could potentially be attributed to a difference in adsorption of collagen, which in turn affects cellular adhesion strength, proliferation and the rate of extracellular matrix remodelling, but the results to support this are insignificant, suggesting a more thorough investigation into this process should be carried out.

An additional in vitro study should be performed in the future in order to evaluate how the adhesion-associated protein expression and activation changes over longer periods of time (several days or a couple of weeks) and whether that impacts cellular adhesion, viability and metabolism in a significant way.

Future studies in tissue engineering are likely to focus on integrating several materials with different properties in order to precisely guide cellular behaviour as well as investigating the role that artificial 3D niches with different geometries might play in affecting cellular differentiation and tissue maturation. As such, an extra option in the form of Al-containing hybrid organometallic polymer, presented in this work, is a choice, judging by its biocompatibility demonstrated in this study. In conjunction with its other advantages, like great 3D structuring capabilities, we believe the materials described in this work have potential to add value for the field of biomedical engineering.

Supplementary Materials: The following are available online at http://www.mdpi.com/1996-1944/12/23/3932/s1, Figure S1: A curve of collagen I content vs Sirius Red absorbance at 550 nm. An average of three independent experiments with n = 3 in each is presented ± standard deviation. Regression analysis was performed, showing that the dependence is linear in the measurement range (0–100 µg) with $R^2 = 0.994$ and $f(x) = 0.0199x - 0.0322$, Figure S2: Characterization of rat skeletal muscle-derived progenitor cells. Fluorescent microscope images of cells immunostained for CD45, CD34, c-kit and Myf5. The scale bars correspond to 100 microns, Figure S3: Representative images of cells grown on glass, Al and Zr surfaces after 24, 48, 72 and 96 h. The scale bars correspond to 100 microns, Figure S4: Flow cytometry data on cellular viability on Glass, Al and Zr surfaces after 24, 48, 72 and 96 h. The results are presented as average absorbance +/- standard deviation, derived from 3 independent experiments with n = 3 per experiment. No statistically significant differences were found between the surfaces, Figure S5: Images of the Western blot membranes, showing the expression levels of p-Akt (Ser473), p-Akt (Thr308), Akt and FAK proteins. Tubulin was used as a reference. From left to right: 1st membrane: ladder, Al(4 h), Al(24 h), Zr(4 h), Zr(24 h), Glass(4 h), Glass(24 h), Al(4 h), Al(24 h), Zr(4 h), Zr(4 h); 2nd membrane: ladder, Zr(24 h), Glass(4 h), Glass(24 h), Al(24h), Zr(4 h), Zr(24 h), Glass(4 h), Glass(24 h).

Author Contributions: Conceptualization, E.B. and D.B.; methodology, E.B., S.J.B., J.G.H., D.B.; formal analysis, all authors; investigation, all authors; resources, J.G.H. and D.B.; data curation, D.B.; writing—original draft preparation, all authors; writing—review and editing, E.B., J.G.H. and D.B.; supervision, J.G.H. and D.B.; project administration, D.B.; funding acquisition, J.G.H. and D.B.

Funding: This research was funded by INFOBALT Lithuania and the World Federation of Scientists. The work was supported by the European Commission via the Marie Skłodowska-Curie research fellowship programme AngioMatTrain (Grant agreement 317304), and a Lancaster University Faculty of Science and Technology Early Career Internal Grant, a Royal Society Research Grant (RG160449), an EPSRC First Grant (EP/R003823/1), and the Research Council of Lithuania (Grant No. SEN-13/2015).

Conflicts of Interest: The authors declare no conflict of interest.

References

1. Cima, L.G.; Vacanti, J.P.; Vacanti, C.; Ingber, D.; Mooney, D.; Langer, R. Tissue Engineering by Cell Transplantation Using Degradable Polymer Substrates. *J. Biomech. Eng.* **1991**, *113*, 143–151. [CrossRef] [PubMed]
2. Langer, R.; Vacanti, J. Advances in tissue engineering. *J. Pediatr. Surg.* **2016**, *51*, 8–12. [CrossRef] [PubMed]
3. Crapo, P.M.; Gilbert, T.W.; Badylak, S.F. An overview of tissue and whole organ decellularization processes. *Biomaterials* **2011**, *32*, 3233–3243. [CrossRef] [PubMed]
4. Melchels, F.P.W.; Domingos, M.A.N.; Klein, T.J.; Malda, J.; Bartolo, P.J.; Hutmacher, D.W. Additive manufacturing of tissues and organs. *Prog. Polym. Sci.* **2012**, *37*, 1079–1104. [CrossRef]
5. Raimondi, M.T.; Eaton, S.M.; Nava, M.M.; Laganà, M.; Cerullo, G.; Osellame, R. Two-photon laser polymerization: From fundamentals to biomedical application in tissue engineering and regenerative medicine. *J. Appl. Biomater. Funct. Mater.* **2012**, *10*, 55–65. [CrossRef]
6. Selimis, A.; Mironov, V.; Farsari, M. Direct laser writing: Principles and materials for scaffold 3D printing. *Microelectron. Eng.* **2014**, *132*, 83–89. [CrossRef]
7. Kufelt, O.; El-Tamer, A.; Sehring, C.; Schlie-Wolter, S.; Chichkov, B.N. Hyaluronic acid based materials for scaffolding via two-photon polymerization. *Biomacromolecules* **2014**, *15*, 650–659. [CrossRef]
8. Ovsianikov, A.; Deiwick, A.; Van Vlierberghe, S.; Dubruel, P.; Möller, L.; Drager, G.; Chichkov, B. Laser fabrication of three-dimensional CAD scaffolds from photosensitive gelatin for applications in tissue engineering. *Biomacromolecules* **2011**, *12*, 851–858. [CrossRef]
9. Ovsianikov, A.; Malinauskas, M.; Schlie, S.; Chichkov, B.; Gittard, S.; Narayan, R.; Löbler, M.; Sternberg, K.; Schmitz, K.P.; Haverich, A. Three-dimensional laser micro- and nano-structuring of acrylated poly(ethylene glycol) materials and evaluation of their cytoxicity for tissue engineering applications. *Acta Biomater.* **2011**, *7*, 967–974. [CrossRef]
10. Thiel, M.; Fischer, J.; Von Freymann, G.; Wegener, M. Direct laser writing of three-dimensional submicron structures using a continuous-wave laser at 532 nm. *Appl. Phys. Lett.* **2010**, *97*, 221102. [CrossRef]
11. Rekštyte, S.; Jonavičius, T.; Malinauskas, M. Direct laser writing of microstructures on optically opaque and reflective surfaces. *Optics Laser Eng.* **2014**, *53*, 90–97. [CrossRef]
12. Rekštytė, S.; Kaziulionytė, E.; Balčiūnas, E.; Kaškelytė, D.; Malinauskas, M. Direct laser fabrication of composite material 3D microstructured scaffolds. *J. Laser Micro Nanoeng.* **2014**, *9*, 25–30. [CrossRef]
13. Liu, X.; Miller, A.L.; Park, S.; George, M.N.; Waletzki, B.E.; Xu, H.; Terzic, A.; Lu, L. Two-Dimensional Black Phosphorus and Graphene Oxide Nanosheets Synergistically Enhance Cell Proliferation and Osteogenesis on 3D Printed Scaffolds. *ACS Appl. Mater. Interfaces* **2019**, *11*, 23558–23572. [CrossRef] [PubMed]
14. Zein, I.; Hutmacher, D.W.; Tan, K.C.; Teoh, S.H. Fused deposition modeling of novel scaffold architectures for tissue engineering applications. *Biomaterials* **2002**, *23*, 1169–1185. [CrossRef]
15. Lin, M.; Firoozi, N.; Tsai, C.T.; Wallace, M.B.; Kang, Y. 3D-printed flexible polymer stents for potential applications in inoperable esophageal malignancies. *Acta Biomater.* **2019**, *83*, 119–129. [CrossRef] [PubMed]
16. Olakanmi, E.O.; Cochrane, R.F.; Dalgarno, K.W. A review on selective laser sintering/melting (SLS/SLM) of aluminium alloy powders: Processing, microstructure, and properties. *Prog. Mater. Sci.* **2015**, *74*, 401–477. [CrossRef]
17. Farsari, M.; Chichkov, B.N. Materials processing: Two-photon fabrication. *Nat. Photonics* **2009**, *3*, 450–452.
18. Wang, X.; Lin, M.; Kang, Y. Engineering Porous β-Tricalcium Phosphate (β-TCP) Scaffolds with Multiple Channels to Promote Cell Migration, Proliferation, and Angiogenesis. *ACS Appl. Mater. Interfaces* **2019**, *11*, 9223–9232. [CrossRef]
19. Reignier, J.; Huneault, M.A. Preparation of interconnected poly(ε(lunate)-caprolactone) porous scaffolds by a combination of polymer and salt particulate leaching. *Polymer (Guildf.)* **2006**, *47*, 4703–4717.
20. Farsari, M.; Vamvakaki, M.; Chichkov, B.N. Multiphoton polymerization of hybrid materials. *J. Opt.* **2010**, *12*, 124001. [CrossRef]
21. Balčiūnas, E.; Baldock, S.J.; Dreižė, N.; Grubliauskaitė, M.; Coultas, S.; Rochester, D.L.; Valius, M.; Hardy, J.G.; Baltriukienė, D. 3D printing hybrid organometallic polymer-based biomaterials via laser two-photon polymerization. *Polym. Int.* **2019**, *68*, 1928–1940. [CrossRef]

22. Malinauskas, M.; Gaidukevičiute, A.; Purlys, V.; Žukauskas, A.; Sakellari, I.; Kabouraki, E.; Candiani, A.; Gray, D.; Pissadakis, S.; Gadonas, R.; et al. Direct laser writing of microoptical structures using a Ge-containing hybrid material. *Metamaterials* **2011**, *5*, 135–140. [CrossRef]
23. Sakellari, I.; Gaidukeviciute, A.; Giakoumaki, A.; Gray, D.; Fotakis, C.; Farsari, M.; Vamvakaki, M.; Reinhardt, C.; Ovsianikov, A.; Chichkov, B.N. Two-photon polymerization of titanium-containing sol-gel composites for three-dimensional structure fabrication. *Appl. Phys. A Mater. Sci. Process.* **2010**, *100*, 359–364. [CrossRef]
24. Kabouraki, E.; Giakoumaki, A.N.; Danilevicius, P.; Gray, D.; Vamvakaki, M.; Farsari, M. Redox multiphoton polymerization for 3D nanofabrication. *Nano Lett.* **2013**, *13*, 3831–3835. [CrossRef] [PubMed]
25. Ovsianikov, A.; Viertl, J.; Chichkov, B.; Oubaha, M.; MacCraith, B.; Sakellari, I.; Giakoumaki, A.; Gray, D.; Vamvakaki, M.; Farsari, M.; et al. Ultra-low shrinkage hybrid photosensitive material for two-photon polymerization microfabrication. *ACS Nano* **2008**, *2*, 2257–2262. [CrossRef] [PubMed]
26. Malinauskas, M.; Baltriukiene, D.; Kraniauskas, A.; Danilevicius, P.; Jarasiene, R.; Sirmenis, R.; Zukauskas, A.; Balciunas, E.; Purlys, V.; Gadonas, R.; et al. In vitro and in vivo biocompatibility study on laser 3D microstructurable polymers. *Appl. Phys. A Mater. Sci. Process.* **2012**, *108*, 751–759. [CrossRef]
27. Käpylä, E.; Sorkio, A.; Teymouri, S.; Lahtonen, K.; Vuori, L.; Valden, M.; Skottman, H.; Kellomäki, M.; Juuti-Uusitalo, K. Ormocomp-Modified glass increases collagen binding and promotes the adherence and maturation of human embryonic stem cell-derived retinal pigment epithelial cells. *Langmuir* **2014**, *30*, 14555–14565. [CrossRef]
28. Thingiverse. Available online: https://www.thingiverse.com (accessed on 24 October 2019).
29. Tullberg-Reinert, H.; Jundt, G. In situ measurement of collagen synthesis by human bone cells with a sirius red-based colorimetric microassay: Effects of transforming growth factor beta2 and ascorbic acid 2-phosphate. *Histochem. Cell Biol.* **1999**, *112*, 271–276. [CrossRef]
30. Mercille, S.; Massie, B. Induction of apoptosis in nutrient-deprived cultures of hybridoma and myeloma cells. *Biotechnol. Bioeng.* **1994**, *44*, 1140–1154. [CrossRef]
31. Raimondi, M.T.; Eaton, S.M.; Laganà, M.; Aprile, V.; Nava, M.M.; Cerullo, G.; Osellame, R. Three-dimensional structural niches engineered via two-photon laser polymerization promote stem cell homing. *Acta Biomater.* **2013**, *9*, 4579–4584. [CrossRef]
32. Murphy, W.L.; McDevitt, T.C.; Engler, A.J. Materials as stem cell regulators. *Nat. Mater.* **2014**, *13*, 547–557. [CrossRef] [PubMed]
33. Lamont, A.C.; Restaino, M.A.; Kim, M.J.; Sochol, R.D. A facile multi-material direct laser writing strategy. *Lab Chip* **2019**, *19*, 2340–2345. [CrossRef] [PubMed]
34. Malinauskas, M.; Rekštyte, S.; Lukoševičius, L.; Butkus, S.; Balčiunas, E.; Pečiukaityte, M.; Baltriukiene, D.; Bukelskiene, V.; Butkevičius, A.; Kucevičius, P.; et al. 3D microporous scaffolds manufactured via combination of fused filament fabrication and direct laser writing ablation. *Micromachines* **2014**, *5*, 839–858. [CrossRef]
35. Ber, S.; Torun Köse, G.; Hasirci, V. Bone tissue engineering on patterned collagen films: An in vitro study. *Biomaterials* **2005**, *26*, 1977–1986. [CrossRef]
36. Ramanathan, G.; Singaravelu, S.; Muthukumar, T.; Thyagarajan, S.; Perumal, P.T.; Sivagnanam, U.T. Design and characterization of 3D hybrid collagen matrixes as a dermal substitute in skin tissue engineering. *Mater. Sci. Eng. C* **2017**, *72*, 359–370. [CrossRef]
37. Antoine, E.E.; Vlachos, P.P.; Rylander, M.N. Review of Collagen I Hydrogels for Bioengineered Tissue Microenvironments: Characterization of Mechanics, Structure, and Transport. *Tissue Eng. Part B Rev.* **2014**, *20*, 683–696. [CrossRef]
38. Sarker, B.; Hum, J.; Nazhat, S.N.; Boccaccini, A.R. Combining collagen and bioactive glasses for bone tissue engineering: A review. *Adv. Healthc. Mater.* **2015**, *4*, 176–194. [CrossRef]
39. Dong, C.; Lv, Y. Application of collagen scaffold in tissue engineering: Recent advances and new perspectives. *Polymers (Basel).* **2016**, *8*, 42. [CrossRef]
40. Ying, P.; Jin, G.; Tao, Z. Competitive adsorption of collagen and bovine serum albumin - Effect of the surface wettability. *Colloids Surfaces B Biointerfaces* **2004**, *33*, 259–263. [CrossRef]
41. Nieponice, A.; Soletti, L.; Guan, J.; Hong, Y.; Gharaibeh, B.; Maul, T.M.; Huard, J.; Wagner, W.R.; Vorp, D.A. In Vivo Assessment of a Tissue-Engineered Vascular Graft Combining a Biodegradable Elastomeric Scaffold and Muscle-Derived Stem Cells in a Rat Model. *Tissue Eng. Part A* **2010**, *16*, 1215–1223. [CrossRef]

42. Shi, N.; Chen, S.Y. Mechanisms simultaneously regulate smooth muscle proliferation and differentiation. *J. Biomed. Res.* **2014**, *28*, 40–46. [PubMed]
43. Reddig, P.J.; Juliano, R.L. Clinging to life: Cell to matrix adhesion and cell survival. *Cancer Metastasis Rev.* **2005**, *24*, 425–439. [CrossRef] [PubMed]
44. Lee, D.Y.; Li, Y.S.J.; Chang, S.F.; Zhou, J.; Ho, H.M.; Chiu, J.J.; Chien, S. Oscillatory flow-induced proliferation of osteoblast-like cells is mediated by $\alpha v \beta 3$ and $\beta 1$ integrins through synergistic interactions of focal adhesion kinase and Shc with phosphatidylinositol 3-kinase and the Akt/mTOR/p70S6K pathway. *J. Biol. Chem.* **2010**, *285*, 30–42. [CrossRef] [PubMed]
45. Martini, M.; De Santis, M.C.; Braccini, L.; Gulluni, F.; Hirsch, E. PI3K/AKT signaling pathway and cancer: An updated review. *Ann. Med.* **2014**, *46*, 372–383. [CrossRef]
46. Lee, S.; Choi, E.; Cha, M.J.; Hwang, K.C. Cell adhesion and long-term survival of transplanted mesenchymal stem cells: A prerequisite for cell therapy. *Oxid. Med. Cell. Longev.* **2015**, *2015*, 1–9. [CrossRef] [PubMed]

© 2019 by the authors. Licensee MDPI, Basel, Switzerland. This article is an open access article distributed under the terms and conditions of the Creative Commons Attribution (CC BY) license (http://creativecommons.org/licenses/by/4.0/).

Article

Lignin: A Biopolymer from Forestry Biomass for Biocomposites and 3D Printing

Mihaela Tanase-Opedal [1], Eduardo Espinosa [2], Alejandro Rodríguez [2] and Gary Chinga-Carrasco [1,*]

[1] RISE PFI, Høgskoleringen 6b, 7491 Trondheim, Norway; mihaela.tanase@rise-pfi.no
[2] Chemical Engineering Department, Faculty of Science, Universidad de Córdoba, Building Marie-Curie, Campus de Rabanales, 14014 Córdoba, Spain; eduardo.espinosa@uco.es (E.E.); a.rodriguez@uco.es (A.R.)
* Correspondence: gary.chinga.carrasco@rise-pfi.no

Received: 8 August 2019; Accepted: 12 September 2019; Published: 16 September 2019

Abstract: Biopolymers from forestry biomass are promising for the sustainable development of new biobased materials. As such, lignin and fiber-based biocomposites are plausible renewable alternatives to petrochemical-based products. In this study, we have obtained lignin from Spruce biomass through a soda pulping process. The lignin was used for manufacturing biocomposite filaments containing 20% and 40% lignin and using polylactic acid (PLA) as matrix material. Dogbones for mechanical testing were 3D printed by fused deposition modelling. The lignin and the corresponding biocomposites were characterized in detail, including thermo-gravimetric analysis (TGA), differential scanning calorimetry (DSC), Fourier transform infrared (FTIR) spectroscopy, X-ray diffraction analysis (XRD), antioxidant capacity, mechanical properties, and scanning electron microscopy (SEM). Although lignin led to a reduction of the tensile strength and modulus, the reduction could be counteracted to some extent by adjusting the 3D printing temperature. The results showed that lignin acted as a nucleating agent and thus led to further crystallization of PLA. The radical scavenging activity of the biocomposites increased to roughly 50% antioxidant potential/cm^2, for the biocomposite containing 40 wt % lignin. The results demonstrate the potential of lignin as a component in biocomposite materials, which we show are adequate for 3D printing operations.

Keywords: lignin; polylactic acid (PLA); 3D printing; biocomposites; biopolymers

---

1. Introduction

Environmental pollution and the increasing awareness of limited resources have been a major driver in finding renewable alternatives to replace traditional fossil-based plastics, with bio-based materials (for ex., biocomposites) derived from carbon-neutral feedstocks [1]. Natural fibres have various advantages, including good mechanical properties, no emission of toxic substances, and reduction in cost [2]. Additionally, natural fiber-reinforced biocomposites have attracted increasing attention due to several beneficial properties, e.g., low-cost, good mechanical properties, and lightweight.

Poly (lactic) acid (PLA) is a promising biopolymer which has been introduced commercially as a renewable alternative to fossil-based polymers [3,4]. A number of promising PLA-based products are presently found commercially, such as automotive parts [5]. PLA provides good mechanical properties and relatively easy melt-processability. However, PLA is relatively expensive and some properties (e.g., brittleness) limit the utilization of PLA in some applications [6]. Therefore, additional polymers can be used in PLA blends in order to tailor the properties of the final products [7]. Moreover, some alternatives have been suggested to improve e.g. the mechanical and thermal properties, including the addition of fibre or filler materials [8] and cellulose nanofibres [9].

Lignin as the second most abundant renewable bio-resource, next to cellulose, is considered as a waste product in several industrial processes. Attempts for lignin valorisation have been published in a vast number of papers and reviews over the last years [10–12].

The notable properties of lignin, such as highly abundance, low-cost and biodegradability, high carbon content, high aromaticity, and reinforcing capability make it a good candidate as a potential component for biocomposites [13]. Each year, over 50 million tons of lignin are produced worldwide as a side product from biorefineries, of which 98% are burned to generate energy. Only 2% of the lignin has been used for other purposes, mainly in applications such as dispersants, adhesives, and fillers [14–17]. The commercial lignin is mainly lignosulfonates from sulphite pulp mills (about 1 mill. tons/year) and less than 100,000 tons/year of kraft lignin [18]. Since lignin has lower energy content than coal and because lignin-rich side streams are wet, the energy value is limited to 50 US Dollar/dry ton [19]. Thus, cost-efficient valorisation of lignin into value-added products offers a significant opportunity to enhance operational efficiency and generates additional revenue so that the production of bioethanol or other products from the hydrolysed carbohydrates becomes more competitive. However, lignin properties (for ex., high heterogeneity and complex structure), make it difficult to predict how the lignin loading will affect the properties of a given biocomposite. Recent reviews summarized the research done on lignin-reinforced thermoplastic biocomposites [20,21].

Lignin can be used as a component in biocomposites, with or without modification, depending on the target application. Gordobil et al. [22] acetylated kraft lignin to improve the affinity with PLA. However, incorporation of kraft lignin decreased the tensile strength properties of the PLA with increasing lignin loading of 10 wt % and 20 wt %. Furthermore, when acetylated lignin was blended with a thermoplastic the tensile strength was improved [21,22]. Without modification, lignin can be directly incorporated into a polymeric matrix, such as UV-light stabilizer, antioxidant, flame retardant, plasticizer, and flow enhancer to reduce production cost, reduce plastic, and potentially improve material properties [23–26].

Lignin can also be used as a coupling agent in natural fibre biocomposites. Lignin can act as a compatibilizer between the hydrophilic fibres and hydrophobic matrix polymer, thus strengthening the fibre matrix interface [27–29]. Lignin treatments of hemp fibres [28] and flax fibres [27] have been shown to improve compatibility between fibres and the thermoset matrix, thus also improving the mechanical properties of the biocomposites. Graupner [29] reported increased tensile properties of compression-moulded PLA-cotton composites when the fibres were treated with lignin.

In our study, lignin was extracted by a soda process. Soda pulping pre-treatment is similar to that of alkaline pulping process, which uses alkali (e.g., NaOH, Ca(OH)$_2$) to solubilize or depolymerize lignin, and make lignin extractable from biomass matrix [30]. Soda pre-treatment disrupts the lignin structure and breaks the aryl-ether, ester, and C-C linkages among the lignin and hemicelluloses and hence open up the lignocellulose structure [31]. This process has mainly been studied on herbaceous biomass and to some extent hardwood [12,32,33]. The main difference of this process compared to the chemical pulping process is the moderate treatment severity, separating lignin with low condensation structure [19]. The soda process for obtaining lignin has some essential advantages from the environmental point of view in comparison with the sulphate process, namely, the soda cooking liquor has a lower content of low molecular products of wood degradation, which get to waste water, and the formation of the unpleasant sulphur organic compound odorants is prevented [34]. Soda lignin has found application in multiple areas: the production of phenolic resins [35,36], animal nutrition [37], dispersants [38], and synthesis of polymers [39]. A major difference between soda lignin compared with kraft lignin and lignosulphonates is that soda lignin is sulphur-free, without odour and with a chemical composition closed to pure lignin [38,39]. Additionally, soda lignin derived from non-wood plants contains a larger fraction of carboxyl groups and p-hydroxyl units [39].

Three-dimensional (3D) printing has evolved rapidly in recent years. This technology allows the production of unique, complex and customized structures by digitized and computer assisted processes, reducing production time and costs [40]. In addition, less waste in production and lower chemical consumption is required by this technique in comparison with the traditional processing manufacturing. Fused deposition modelling (FDM) is one of the most used 3D printing technologies, which consists in the melting of thermoplastic materials at high temperature, which are solidified when cooling. Currently, there is great interest in the use of biomass and biomass components for use in 3D printing by FDM. However, unlike petroleum-derived thermoplastic compounds, lignocellulosic components are difficult to melt for the extrusion and injection moulding processes. Therefore, the development of new materials from biomass suitable for 3D printing is a challenge [41–43]. The production of pure lignin composites is limited by its high thermal transition temperature and high flow resistance [40]. For this reason, lignin is mixed with other polymers that favour its melting behaviour and flow. Recently, organosolv hardwood lignin [40] and kraft softwood lignin [44] have been applied to manufacture filaments for FDM, based on acrylonitrile-butadiene-styrene and PLA polymers, respectively.

The purpose of this study was to demonstrate the suitability of PLA/soda lignin biocomposite filament for 3D printing. A motivation for selecting soda lignin is that it is sulphur-free. Soda lignin was thus expected to reduce the typical smell that is experienced when melt-processing biocompounds containing kraft lignin or lignosulfonates. Thus, samples with varying PLA/soda lignin weight ratios were manufactured and the mechanical (tensile testing), thermal (TGA, DSC analysis), morphological (SEM), FTIR, X-ray diffraction, and antioxidant properties were assessed.

2. Materials and Methods

2.1. Raw Materials

The lignin used in this study is a softwood lignin extracted from cooking liquor, using a soda cooking process of Norway spruce chips, collected from Norske Skog, Skogn, Norway. The PLA used in this study was a commercial grade for 3D printing (Ingeo PLA 3D850, NatureWorks LCC, Minnetonka, MN, USA).

2.2. Lignin Extraction

Lignin was extracted from Norway spruce by soda process using a MK circulation reactor. In this method, 400 g dried chips were cooked with cooking liquor which comprised 30% NaOH (30 g NaOH/100 g dried chips). The liquid wood ratio was 7.5:1. The temperature was increased slowly in order to get a good impregnation of the chips with the cooking liquor. Once the reactor reached the operating temperature of 180 °C, it was maintained for 100 min. After cooling, the cooking liquor was removed from the bottom of the reactor.

2.3. Lignin Precipitation

Sulphuric acid was slowly added to the cooking liquor under agitation until final pH of 2.5. A color change from black to brown was observed at pH 5.5. In addition, a viscosity change was observed, the liquid was more viscose at lower pH. These changes occur due to the initial stage of the lignin precipitation. The mixture was then centrifuged (3500 rpm, 10 min) to recover the lignin. The lignin was repeatedly washed with water, then dried in an oven at 105 °C overnight. Volatile organic compounds (VOC) were removed after drying in a 2L Parr reactor (N4622, Parr Instruments, Illinois, USA).

2.4. Filaments and 3D Printing

Neat PLA (100%) and blends of PLA and lignin (20 wt % and 40 wt %) were extruded through a Noztek Xcalibur (Table 1). The temperatures in the three heating chambers were 200 °C, 205 °C, and 210 °C. The speed of the screw extruder was set to 15 mm/s. The target diameter of the filaments was 1.75 mm. The filaments were extruded twice in order to improve the mixing of the PLA and lignin.

Table 1. Mechanical properties of PLA and PLA/Lignin biocomposites.

3D Printed Sample	E_t (MPA)	σ_M (MPa)	ε_M %
PLA	2890 ± 14.14	58.45 ± 0.55	2.45 ± 0.10
PLA + 20%Lignin	2460 ± 155.56	39.35 ± 1.05	1.8 ± 0.10
PLA + 40%Lignin (205 °C)	1955 ± 19.92	32 ± 2.10	1.8 ± 0.20
PLA + 40%Lignin (215 °C)	2695 ± 148.49	45.65 ± 0.05	1.9 ± 0.08
PLA + 40%Lignin (230 °C)	1930 ± 183.85	29.25 ± 1.35	1.65 ± 0.10

E_t (tensile elastic modulus); σ_M (tensile strength); ε_M (elongation at tensile).

An original Prusa i3 MK3S was used for 3D printing by fused deposition modeling (FDM), using a 0.4 mm nozzle and a printing speed of 35 mm/s. Dogbone samples (length 63 mm, width 3 mm) were printed for mechanical testing. Three sets of dogbones were printed with varying printing temperatures of 205, 215, and 230 °C. The printing direction was 45°. In addition, a smartphone protective case was printed as a demonstration.

2.5. Characterisation

The lignin and carbohydrates analyses were performed according to the standard procedures of NREL (NREL/TP–510–42623). The lignin samples were hydrolysed in two steps using sulfuric acid: first step hydrolysis was done with 72 wt % H_2SO_4, 30 °C for 60 min, followed by the second step with 4 wt % H_2SO_4 at 121 °C for 1 h. The resulting supernatant was filtered (0.2uL) and the filtrates were analysed by High Performance Anion Exchange Chromatography (HPAEC, ICS-5000, Dionex, CarboPac PA1 4 × 250 mm column), for carbohydrates. Klasson lignin (acid insoluble lignin) contents were determined gravimetrically, while acid soluble lignin was determined by using TAPPI method (TAPPI UM 250). The absorbance was measured with UV-VIS spectrometry at 205 nm, ε of 110. Carbohydrates were detected on a pulsed amperometry detector (PAD). All carbohydrate contents are reported as anhydrosugars. Purity of lignin samples was calculated from the sum of the ash and sugar results.

2.5.1. Thermo-Gravimetric Analysis (TGA) and Differential Scanning Calorimetry (DSC)

TGA was used to determine the thermal stability, decomposition temperature, and char yield for soda lignin, PLA, and different blends of lignin/PLA. The analyses were performed using a Netzsch Jupiter F3 equipment, operating in nitrogen environment. Samples for TGA for each measurement were maintained at 14 ± 5 mg and scans were preformed from 30 °C to 800 °C with the heating rate of 10 °C/min to observe thermal degradation and stability of lignin, PLA, and the corresponding biocomposites. DSC was performed to measure the glass transition temperature (Tg) of soda lignin.

2.5.2. Scanning Electron Microscopy (SEM)

The fracture surface of the dogbones after mechanical testing was sputtered with a layer of gold to make it conductive under the electron beam. Scanning electron microscopy (SEM, Hitachi SU3500, Hitachi High-Technologies Co., Tokyo, Japan) was performed in secondary electron imaging mode using an acceleration voltage of 5 kV and a working distance of 5–6 mm.

2.5.3. Mechanical Testing

Test specimens obtained from the 3D printing process were used for tensile testing of the biocomposites.

The 3D printed dogbones were mechanically tested with a Zwick Roell Proline (Zwick GmbH & Co. KG, Ulm, Germany) and a load cell of 2.5 kN. Four specimens of each series were tested. The speed and the grip distance were 20 mm/min and 50 mm, respectively.

2.5.4. Fourier Transform Infrared (FTIR) Spectroscopy

FTIR spectra were collected using a spectrometer FTIR-ATR Perkin Elmer Spectrum (Perkin Elmer, Waltham, United States). Two single spectra were collected in the wavelength range from 4000 to 450 cm^{-1} with a resolution of 4 cm^{-1} and a total of 40 scans.

2.5.5. X-ray Diffraction Analysis (XRD)

X-ray diffraction analysis of the test specimens was carried out in a Bruker D8 Discover Instrument (Bruker Corporation, Karlsruhe, Germany) with a monochromatic source (CuKα1) over an angular range of 5–50° at a scan speed of 1.56°/min.

2.5.6. Antioxidant Activity

The 2,2'-Azino-Bis-3-Ethylbenzothiazoline-6-Sulfonic Acid (ABTS) assay was used to measure the antioxidant activity of the biocomposites. A radical solution (7 mM ABTS and 2.45 mM potassium persulphate) was prepared and left in dark during 14–16 h before testing. The radical solution was adjusted to an absorbance of 0.70 ± 0.02 at 734 nm diluting with ethanol. A specimen of 0.5–1 cm^2 of the biocomposites was added to a 4 mL of radical solution and the absorbance of the solution was measured at 734 nm using ethanol as blank. The absorbance was measured after 6 min. The antioxidant activity was determined according to the following equation [45]:

$$AOP\,(\%) = \frac{A_{734,ABTS6'} - A_{734,sample6'}}{A_{734,ABTS0'}} \times 100 \qquad (1)$$

where $A_{734,ABTS6'}$ is the absorbance at 734 nm of the radical solution after 6 min, $A_{734,sample6'}$ the absorbance at 734 nm of the sample after 6 min, and $A_{734,ABTS0'}$ the absorbance at 734 nm of the radical solution before the 6 min. Because the radical scavenging activity was performed in the surface of the biocomposites, the antioxidant activity was related to the surface exposed to the radical solution, expressed as antioxidant potential, (AOP%/cm^2).

3. Results and Discussion

3.1. Lignin Composition

The yield of lignin extraction was estimated to be 25%, based on Klasson lignin. The chemical composition shows that lignin is composed of 82% acid insoluble lignin, 6% acid soluble lignin, 1.6% carbohydrates, and ash. Surprisingly, glucomannan, the predominant constituent of softwood hemicellulose, was not found in the soda lignin. However, xylan was found in relatively high amounts. This shows that cleavage of the lignin–carbohydrates complexes is less complete for the soda process applied in this study. These carbohydrates may originate from lignin–carbohydrates complexes or from carbohydrates that are trapped during the lignin precipitation step which end up non-covalently bonds in the lignin after drying [46]. Furthermore, the high ash content in soda lignin may partly originate from the salts resulted after neutralization of cooking liquor during precipitation. These results are in accordance with literature results [47]. The chemical composition of lignin and ash content depends on the feedstock as well as on the isolation process. In general, alkaline lignin contains more residual sugar than other type of lignins [47].

The moisture content of isolated lignin was less than 2.5%. Sameni et al. [48] reported that lignin with higher impurities, containing hydrophilic compounds, led to more moisture content.

3.2. TGA

TGA curves reveal the residual mass of materials with respect to the temperature of thermal degradation and was used to assess the thermal stability of the both polymers PLA and Lignin but also of the biocomposites, PLA/20%Lignin and PLA/40%Lignin. The different chemical bonds present in the lignin molecular structure, leads to a range of degradation temperatures, extending from 100 to 800 °C [49]. The results show that 42 wt % of lignin sample still remained at 800 °C. This is due to the formation of highly condensed aromatic structures which have the ability to form char (Figure 1). Degradation of lignin sample can be divided into three stages [36]. Firstly, an initial weight loss between 50–120 °C, due to water evaporation. Secondly, above 220 °C degradation of carbohydrates occurs, which are converted to gasses such as CO, CO_2, and CH_4 [50]. The last stage occurs to around 420 °C, and then continues to lose mass at a slower rate, leading to the formation of gaseous products and phenolics, alcohols and aldehyde acids [50]. PLA, PLA/20%Lignin, and PLA/40%Lignin present a similar thermal degradation behavior. They show a faster thermal degradation between 340–400 °C. However, the residual mass at 800 °C is higher for biocomposites with lignin due to the carbonaceous composition of the lignin. With respect to the starting temperature of decomposition, the lignin shows a faster degradation (326 °C) compared to neat PLA (348 °C). As expected, the biocomposites resulting from the mixtures PLA/Lignin present an intermediate degradation start temperature (347 °C and 339 °C for 20% and 40% lignin content, respectively).

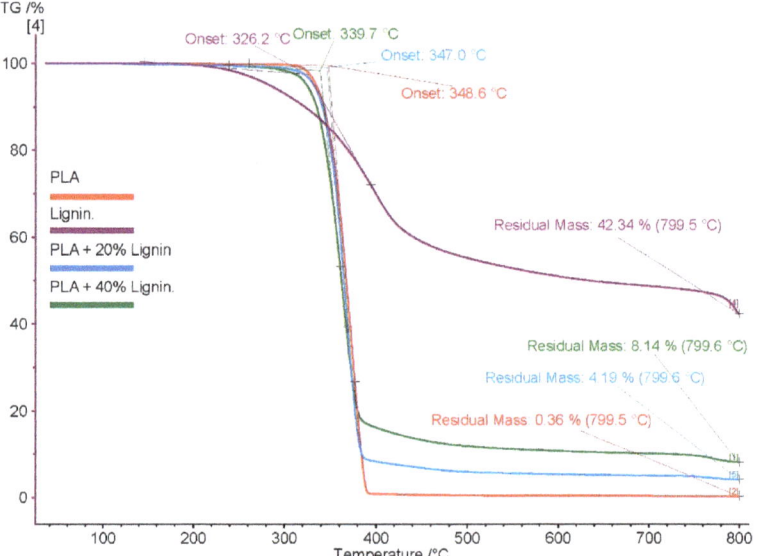

Figure 1. Thermo-gravimetric analysis (TGA) plot of Lignin, polylactic acid (PLA), and PLA/Lignin biocomposites.

3.3. DSC

The reaction energy was measured on soda lignin, polymer and blends of lignin and polymer to assess the thermal degradation. Enthalpy measurements were similar for PLA/40%Lignin (27 J/g) compared with neat PLA (31 J/g). At higher a temperature of 376 °C (PLA) the enthalpy of 628 J/g was similar with PLA/20%Lignin and higher than PLA/40%Lignin (362 °C, 409 J/g), see Figure 2. No significant variations in Tm (temperature maximum degradation) were found with the addition of lignin to PLA matrix. The temperature of maximum degradation occurred between 360–400 °C for all biocomposite samples. Decomposition of aromatic rings is expected above 500 °C [36]. The glass transition temperature (Tg) for soda lignin in our study was found to be at 109 °C. Additionally, the Tg value can be correlated with the molecular weight of lignin [51]. This value is in accordance with the values reported in the literature. The Tg for neat PLA is found to be at 71 °C in this study. The glass transition temperature of the PLA/Lignin biocomposites showed significant shift of the Tg from 71 °C towards lower temperature, Tg of 59 °C for PLA/Lignin 20%. This can be explained by different molecular factors such as interchain hydrogen bonding, crosslinking density, rigid phenyl groups, and molecular mass [52].

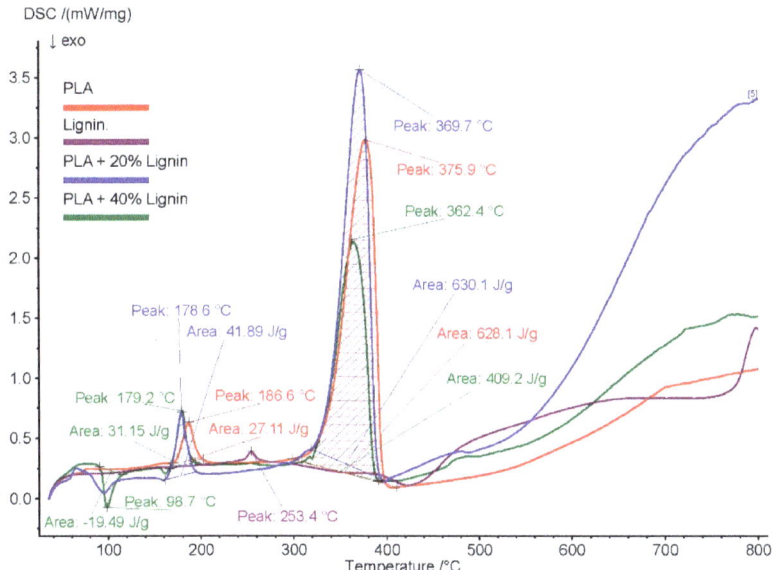

Figure 2. Differential scanning calorimetry (DSC) thermographs of Lignin, PLA, and PLA/lignin biocomposites.

3.4. Mechanical Properties

The filaments (PLA, PLA/20%Lignin, and PLA/40%Lignin) were used to 3D print dogbones for further characterization (Figure 3). The stress–strain curves of the biocomposites with unmodified lignin showed that the mechanical strength decreased when the lignin content increased (Figure 4), which is in accordance with previous studies [21,22]. This is most probably due to a low fusing between the printing layers, especially in biocomposites with 40 wt % of lignin. However, when the printing temperature was increased to 215 °C, the biocomposite revealed a relative increase of the mechanical properties. This was presumptively due to an improved adhesion of the 3D printed layers. The strength–strain curves also show that lignin led to a more fragile biocomposite that resists less deformation before rupture. Consequently, the elastic modulus decreased by 25%–32%, compared to the neat PLA sample. The addition of low content of lignin in PLA biocomposites presented an improvement of the ductility. However, lignin content above 10% has caused a decrease in the plasticity of biocomposites, either for acetylated or unmodified lignin [53].

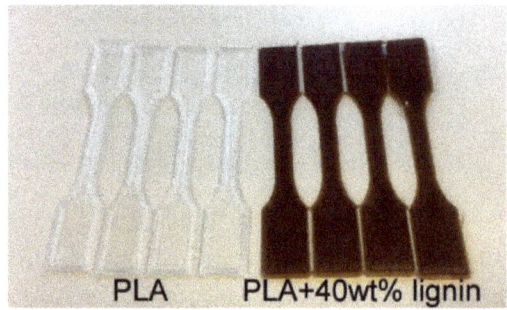

Figure 3. 3D printed dogbones for mechanical testing.

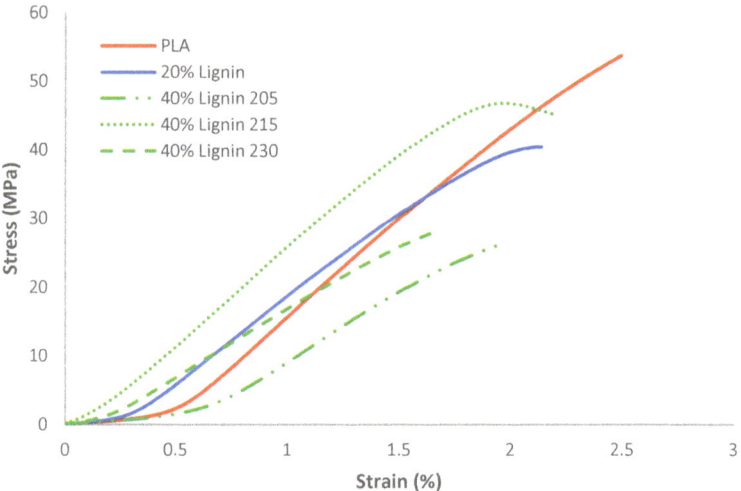

Figure 4. Stress–strain curves for the different biocomposites

3.5. Scanning Electron Microscopy (SEM)

The SEM images (Figure 5) of the fractured surface of the biocomposites provided an insight into the bonding interaction between lignin and PLA. The analysis of the PLA/40% Lignin sample revealed a 3D printed structure where the printed threads are clearly visible, thus confirming a poor inter-layer adhesion and a corresponding low mechanical performance (Table 1). In order to find a suitable printing temperature of the lignin-containing biocomposite and thus increase the inter-layer bonding, two additional temperatures were tested during the 3D printing process (215 °C and 230 °C). The results revealed that a suitable printing temperature for the biocomposite filaments developed in this study was 215 °C (Table 1 and Figure 4). The SEM pictures showed the improvement of the bonding of the printed layers, which favored the mechanical properties of the samples. However, increasing the printing temperature to 230 °C led to a decrease of the tensile strength and modulus. This was potentially caused by the degradation of the carbohydrates in the lignin fraction, which usually occurs over 220 °C, becoming volatile gases and presumptively creating microstructures within the polymeric matrix. Our results are in accordance with literature results of Thakur et al. [20] and Watkins et al. [50].

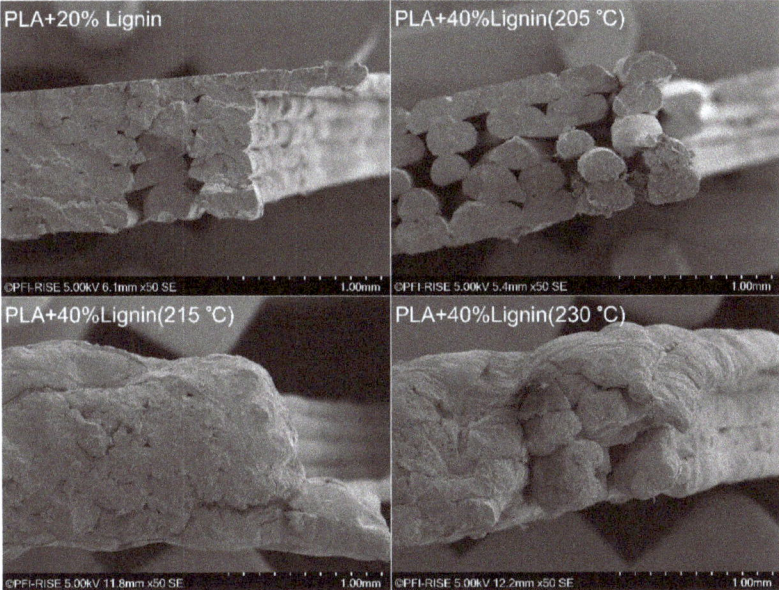

Figure 5. Scanning electron microscopy (SEM) analysis of the fracture surface of tensile tested dogbones.

3.6. FT-IR Spectra and XRD Analysis

FT-IR spectroscopy was applied to assess the functional groups of PLA/Lignin biocomposites at different temperatures. Generally, the curves of PLA/lignin showed similar bands with increased emissivity compared to neat PLA (Figure 6). Neat PLA showed peaks around 2995 cm^{-1} and 2930 cm^{-1}, which are associated to the asymmetric and symmetric stretching vibration of CH$_3$ group. The intense peak at 1749 cm^{-1} is attributed to the C=O stretching vibration [54]. The peak at 1450 cm^{-1} corresponds to CH$_3$ anti-symmetric bending vibration. The peaks at 1385 cm^{-1}, 1360 cm^{-1}, 1316 cm^{-1}, and 1300 cm^{-1} are associated to the deformation, symmetric, and bending mode of the CH group, respectively. The peaks at 1182 cm^{-1}, 1084 cm^{-1}, and 1038 cm^{-1} are attributed to C-O-C stretching vibrations [55]. This peak is obviously higher in the PLA/40%Lignin curve, which indicates that addition of lignin led to a higher content of hydroxy groups. In addition, the biocomposites containing lignin showed a small peak at 1510 cm^{-1} due to the C=C groups of the aromatic rings of lignin.

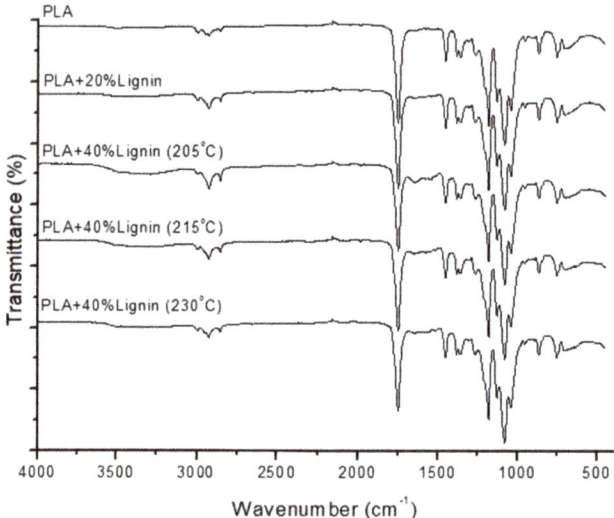

Figure 6. FT-IR spectra of PLA and PLA/lignin biocomposites.

X-ray analysis revealed a change in crystallinity as the lignin was included in the formulation (Figure 7). PLA exhibits a broad peak at 2θ degrees = 10°–25° associated with the semicrystalline nature of PLA. The appearance of peaks at 2θ = 32° and 34.5° in lignin-containing biocomposites indicated further crystallization of PLA, due to the action of lignin as nucleating agent [56].

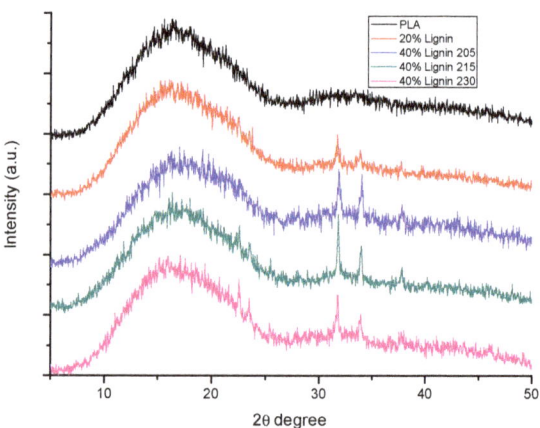

Figure 7. X-ray diffraction analysis (XRD) patterns of PLA and PLA/Lignin biocomposites.

3.7. Antioxidant Properties

Antioxidant capacity of the biocomposites was measured and expressed as radical scavenging activity (RSA). PLA shows a low antioxidant (activity only associated with the surface interaction with the radical ABTS) (Figure 8). However, the lignin-containing biocomposites show a significantly higher radical scavenging activity (~50%) due to the antioxidant activity of lignin. Although, the biocomposites containing 40% lignin have a slightly higher RSA compared to the PLA/20% Lignin, the differences are not significant. The reaction of ABTS with the biocomposite material occurs mostly on the surface of the specimens, which may explain this behavior. The antioxidant activity of the lignin has been reported previously in several studies [45,57,58], including the blending of low amounts of kraft lignin (0.5 wt %–3 wt %) in PLA for potential biomedical applications [44]. Materials with high antioxidant activity are highly demanded for application in food packaging and biomedicine. A wide variety of antioxidant compounds are described in literature (resveratrol, curcumin, ascorbic acid, carotenoids, etc.), however, these compounds are generally expensive. For this reason, the use of lignin in biocomposites is proposed as a low-cost option to produce materials with high antioxidant capacity.

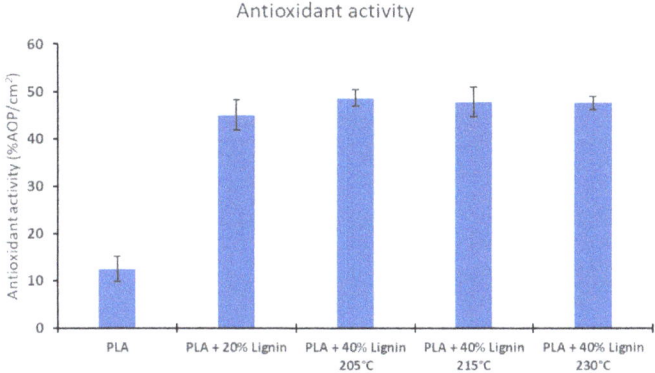

Figure 8. Antioxidant activity of PLA and biocomposites.

Keep in mind that lignin is a natural antioxidant, and lignin has been proposed to stabilize a given material against photo- and thermo-oxidation [14–17,23]. The antioxidant property of soda lignin has been confirmed in this study where the printed materials containing lignin has a significant larger antioxidant capability compared to PLA (Figure 7). This suggests that the biocomposites developed in this study are also suitable for additional materials, e.g., food packaging applications.

The suitability of the PLA/lignin biocomposite filament for 3D printing was also tested, by printing a smartphone protective case (Figure 9). The printing process revealed a good performance of the lignin-containing filament, and a functional protective case was effectively 3D printed. PLA/Lignin filaments are a plausible option for lignin utilization with potential in, e.g., rapid prototyping and consumer products [59]. It is worth to mention that the typical smell from some lignins was not detected during the extrusion of the filaments or during the printing process, which is an additional advantage of using soda lignin in PLA biocomposite materials.

Figure 9. 3D printing of a smartphone protective case with PLA/lignin biocomposite filament.

4. Conclusions

PLA/lignin biocomposites with different lignin loading were prepared and characterized in detail. Thermogravimetric analysis indicated that the thermal decomposition of lignin occurred over a wide temperature range. PLA/lignin biocomposites showed the highest antioxidant activity, due to the antioxidant activity of lignin. Biocomposites exhibited good extrudability and flowability with no observable agglomeration of the lignin. This suggests that lignin-containing biocomposites are plausible alternatives for 3D printing applications.

Author Contributions: Idea and supervision: G.C.-C.; Investigation: M.T.-O., E.E., G.C.-C., Draft preparation, reviewing and editing: M.T.-O., E.E., A.R., G.C.-C.

Funding: Part of this work was funded by the Research Council of Norway through the ALLOC project (grant 282310). The authors thank the COST Action LignoCOST (CA17128) for funding the short-term scientific mission of E.E. at RISE PFI and Spain's DGICyT, MICINN for supporting this research within the framework of the Projects CTQ2016-78729-R and the Spanish Ministry of Science and Education through the National Program FPU (Grant Number FPU14/02278).

Acknowledgments: Johnny Kvakland Melbø, Kenneth Aasarød, Ingebjørg Leirset and Cornelis van der Wijst at RISE PFI are acknowledged for valuable assistance in the laboratory work.

Conflicts of Interest: The authors declare no conflict of interest.

References

1. Calvo-Flores, F.G.; Dobado, J.; Isac-García, J.; Martin-Martinez, F. Applications of Modified and Unmodified Lignins. In *Lignin and Lignans as Renewable Raw Material*; John Wiley & Sons Ltd.: New York, NY, USA, 2015; pp. 247–288.
2. Oksman, K.; Skrifvars, M.; Selin, J.F. Natural fibres as reinforcement in polylactic acid (PLA) composites. *Compos. Sci. Technol.* **2003**, *63*, 1317–1324. [CrossRef]
3. Lim, L.T.; Auras, R.; Rubino, M. Processing technologies for poly(lactic acid). *Prog. Polym. Sci.* **2008**, *33*, 820–852. [CrossRef]
4. Ray, S.S.; Yamada, K.; Okamoto, M.; Ueda, K. Polylactide-Layered Silicate Nanocomposite: A Novel Biodegradable Material. *Nano Lett.* **2002**, *2*, 1093–1096. [CrossRef]
5. Pang, X.; Zhuang, X.; Tang, Z.; Chen, X. Polylactic acid (PLA): Research, development and industrialization. *Biotechnol. J.* **2010**, *5*, 1125–1136. [CrossRef] [PubMed]
6. Cohn, D.; Salomon, A.H. Designing biodegradable multiblock PCL/PLA thermoplastic elastomers. *Biomaterials* **2005**, *26*, 2297–2305. [CrossRef] [PubMed]
7. Nofar, M.; Sacligil, D.; Carreau, P.J.; Kamal, M.R.; Heuzey, M.-C. Poly (lactic acid) blends: Processing, properties and applications. *Int. J. Biol. Macromol.* **2019**, *125*, 307–360. [CrossRef]
8. Ochi, S. Mechanical properties of kenaf fibers and kenaf/PLA composites. *Mech. Mater.* **2008**, *40*, 446–452. [CrossRef]
9. Vatansever, E.; Arslan, D.; Nofar, M. Polylactide cellulose-based nanocomposites. *Int. J. Biol. Macromol.* **2019**, *137*, 912–938. [CrossRef]
10. Rinaldi, R.; Jastrzebski, R.; Clough, M.T.; Ralph, J.; Kennema, M.; Bruijnincx, P.C.A.; Weckhuysen, B.M. Paving the Way for Lignin Valorisation: Recent Advances in Bioengineering, Biorefining and Catalysis. *Angew. Chem. Int. Ed.* **2016**, *55*, 8164–8215. [CrossRef]
11. Figueiredo, P.; Lintinen, K.; Hirvonen, J.T.; Kostiainen, M.A.; Santos, H.A. Properties and chemical modifications of lignin: Towards lignin-based nanomaterials for biomedical applications. *Prog. Mater. Sci.* **2018**, *93*, 233–269. [CrossRef]
12. Schutyser, W.; Renders, T.; Van den Bosch, S.; Koelewijn, S.F.; Beckham, G.T.; Sels, B.F. Chemicals from lignin: An interplay of lignocellulose fractionation, depolymerisation, and upgrading. *Chem. Soc. Rev.* **2018**, *47*, 852–908. [CrossRef] [PubMed]
13. Liu, L.; Qian, M.; Song, P.A.; Huang, G.; Yu, Y.; Fu, S. Fabrication of Green Lignin-based Flame Retardants for Enhancing the Thermal and Fire Retardancy Properties of Polypropylene/Wood Composites. *ACS Sustain. Chem. Eng.* **2016**, *4*, 2422–2431. [CrossRef]

14. Zhao, W.; Simmons, B.; Singh, S.; Ragauskas, A.; Cheng, G. From lignin association to nano-/micro-particle preparation: Extracting higher value of lignin. *Green Chem.* **2016**, *18*, 5693–5700. [CrossRef]
15. Grossman, A.; Vermerris, W. Lignin-based polymers and nanomaterials. *Curr. Opin. Biotechnol.* **2019**, *56*, 112–120. [CrossRef] [PubMed]
16. Gellerstedt, G.; Tomani, P.; Axegard, P.; Backlund, B. Lignin recovery and lignin-based products. In *Integrated Forest Biorefineries: Challenges and Opportunities*; RSC Green Chemistry: Cambridge, UK, 2013; Chapter 8; pp. 180–210.
17. Ragauskas, A.J.; Beckham, G.T.; Biddy, M.J.; Chandra, R.; Chen, F.; Davis, M.F.; Davison, B.H.; Dixon, R.A.; Gilna, P.; Keller, M.; et al. Lignin Valorization: Improving Lignin Processing in the Biorefinery. *Science* **2014**, *344*, 1246843. [CrossRef]
18. Christopher, L.P. Integrated Forest Biorefineries: Current State and Development Potential. In *Integrated Forest Biorefineries: Challenges and Opportunities*; RCS Green Chemistry: Cambridge, UK, 2013; Chapter 1; pp. 1–66.
19. Wang, H.; Pu, Y.; Ragauskas, A.; Yang, B. From lignin to valuable products–strategies, challenges, and prospects. *Bioresour. Technol.* **2019**, *271*, 449–461. [CrossRef] [PubMed]
20. Thakur, V.K.; Thakur, M.K.; Raghavan, P.; Kessler, M.R. Progress in Green Polymer Composites from Lignin for Multifunctional Applications: A Review. *ACS Sustain. Chem. Eng.* **2014**, *2*, 1072–1092. [CrossRef]
21. Kun, D.; Pukánszky, B. Polymer/lignin blends: Interactions, properties, applications. *Eur. Polym. J.* **2017**, *93*, 618–641. [CrossRef]
22. Gordobil, O.; Delucis, R.; Egüés, I.; Labidi, J. Kraft lignin as filler in PLA to improve ductility and thermal properties. *Ind. Crop. Prod.* **2015**, *72*, 46–53. [CrossRef]
23. Pouteau, C.; Dole, P.; Cathala, B.; Averous, L.; Boquillon, N. Antioxidant properties of lignin in polypropylene. *Polym. Degrad. Stab.* **2003**, *81*, 9–18. [CrossRef]
24. Domenek, S. Potential of Lignins as Antioxidant Additive in Active Biodegradable Packaging Materials. *J. Polym. Environ.* **2013**, *21*, 692–701. [CrossRef]
25. Mishra, S.B.; Mishra, A.K.; Kaushik, N.K.; Khan, M.A. Study of performance properties of lignin-based polyblends with polyvinyl chloride. *J. Mater. Process. Technol.* **2007**, *183*, 273–276. [CrossRef]
26. De Chirico, A.; Armanini, M.; Chini, P.; Cioccolo, G.; Provasoli, F.; Audisio, G. Flame retardants for polypropylene based on lignin. *Polym. Degrad. Stab.* **2003**, *79*, 139–145. [CrossRef]
27. Thielemans, W.; Can, E.; Morye, S.S.; Wool, R.P. Novel applications of lignin in composite materials. *J. Appl. Polym. Sci.* **2002**, *83*, 323–331. [CrossRef]
28. Thielemans, W.; Wool, R.P. Kraft lignin as fiber treatment for natural fiber-reinforced composites. *Polym. Compos.* **2005**, *26*, 695–705. [CrossRef]
29. Graupner, N. Application of lignin as natural adhesion promoter in cotton fibre-reinforced poly(lactic acid) (PLA) composites. *J. Mater. Sci.* **2008**, *43*, 5222–5229. [CrossRef]
30. Kim, J.S.; Lee, Y.Y.; Kim, T.H. A review on alkaline pretreatment technology for bioconversion of lignocellulosic biomass. *Bioresour. Technol.* **2016**, *199*, 42–48. [CrossRef] [PubMed]
31. Saratale, G.D.; Oh, M.-K. Improving alkaline pretreatment method for preparation of whole rice waste biomass feedstock and bioethanol production. *RSC Adv.* **2015**, *5*, 97171–97179. [CrossRef]
32. González-García, S.; Moreira, M.T.; Artal, G.; Maldonado, L.; Feijoo, G. Environmental impact assessment of non-wood based pulp production by soda-anthraquinone pulping process. *J. Clean. Prod.* **2010**, *18*, 137–145. [CrossRef]
33. Rodríguez, A.; Sánchez, R.; Requejo, A.; Ferrer, A. Feasibility of rice straw as a raw material for the production of soda cellulose pulp. *J. Clean. Prod.* **2010**, *18*, 1084–1091. [CrossRef]
34. Vishtal, A.G.; Kraslawski, A. Challenges in industrial applications of technical lignins. *BioResources* **2011**, *6*, 3547–3568.
35. Gosselink, R.J.A.; Abächerli, A.; Semke, H.; Malherbe, R.; Käuper, P.; Nadif, A.; van Dam, J.E.G. Analytical protocols for characterisation of sulphur-free lignin. *Ind. Crop. Prod.* **2004**, *19*, 271–281. [CrossRef]
36. Tejado, A.; Peña, C.; Labidi, J.; Echeverria, J.M.; Mondragon, I. Physico-chemical characterization of lignins from different sources for use in phenol–formaldehyde resin synthesis. *Bioresour. Technol.* **2007**, *98*, 1655–1663. [CrossRef] [PubMed]
37. Baurhoo, B.; Ruiz-Feria, C.A.; Zhao, X. Purified lignin: Nutritional and health impacts on farm animals—A review. *Anim. Feed Sci. Technol.* **2008**, *144*, 175–184. [CrossRef]

38. Nadif, A.; Hunkeler, D.; Käuper, P. Sulfur-free lignins from alkaline pulping tested in mortar for use as mortar additives. *Bioresour. Technol.* **2002**, *84*, 49–55. [CrossRef]
39. Wörmeyer, K.; Ingram, T.; Saake, B.; Brunner, G.; Smirnova, I. Comparison of different pretreatment methods for lignocellulosic materials. Part II: Influence of pretreatment on the properties of rye straw lignin. *Bioresour. Technol.* **2011**, *102*, 4157–4164. [CrossRef]
40. Nguyen, N.A.; Bowland, C.C.; Naskar, A.K. A general method to improve 3D-printability and inter-layer adhesion in lignin-based composites. *Appl. Mater. Today* **2018**, *12*, 138–152. [CrossRef]
41. Zhao, D.X.; Cai, X.; Shou, G.Z.; Gu, Y.Q.; Wang, P.X. Study on the Preparation of Bamboo Plastic Composite Intend for Additive Manufacturing. *Key Eng. Mater.* **2016**, *667*, 250–258. [CrossRef]
42. Henke, K.; Treml, S. Wood based bulk material in 3D printing processes for applications in construction. *Eur. J. Wood Wood Prod.* **2013**, *71*, 139–141. [CrossRef]
43. Le Duigou, A.; Castro, M.; Bevan, R.; Martin, N. 3D printing of wood fibre biocomposites: From mechanical to actuation functionality. *Mater. Des.* **2016**, *96*, 106–114. [CrossRef]
44. Domínguez-Robles, J.; Martin, N.K.; Fong, M.L.; Stewart, S.A.; Irwin, N.J.; Rial-Hermida, M.I.; Donnelly, R.F.; Larrañeta, E. Antioxidant PLA Composites Containing Lignin for 3D Printing Applications: A Potential Material for Healthcare Applications. *Pharmaceutics* **2019**, *11*, 165. [CrossRef] [PubMed]
45. García, A.; González Alriols, M.; Spigno, G.; Labidi, J. Lignin as natural radical scavenger. Effect of the obtaining and purification processes on the antioxidant behaviour of lignin. *Biochem. Eng. J.* **2012**, *67*, 173–185. [CrossRef]
46. Huijgen, W.J.J.; Telysheva, G.; Arshanitsa, A.; Gosselink, R.J.A.; de Wild, P.J. Characteristics of wheat straw lignins from ethanol-based organosolv treatment. *Ind. Crop. Prod.* **2014**, *59*, 85–95. [CrossRef]
47. Constant, S.; Wienk, H.L.J.; Frissen, A.E.; de Peinder, P.; Boelens, R.; van Es, D.S.; Grisel, R.J.H.; Weckhuysen, B.M.; Huijgen, W.J.J.; Gosselink, R.J.A.; et al. New insights into the structure and composition of technical lignins: A comparative characterisation study. *Green Chem.* **2016**, *18*, 2651–2665. [CrossRef]
48. Sameni, J.; Krigstin, S.; de Santos Rosa, D.; Leao, A.; Sain, M. Thermal Characteristics of Lignin Residue from Industrial Processes. *BioResources* **2013**, *9*, 725–737. [CrossRef]
49. Yang, H.; Yan, R.; Chen, H.; Lee, D.H.; Zheng, C. Characteristics of hemicellulose, cellulose and lignin pyrolysis. *Fuel* **2007**, *86*, 1781–1788. [CrossRef]
50. Watkins, D.; Nuruddin, M.; Hosur, M.; Tcherbi-Narteh, A.; Jeelani, S. Extraction and characterization of lignin from different biomass resources. *J. Mater. Res. Technol.* **2015**, *4*, 26–32. [CrossRef]
51. Schmidl, G. Molecular Weight Characterization and Rheology of Lignins for Carbon Fibers. Ph.D. Thesis, University of Florida, Gainesville, FL, USA, 1992.
52. Heitner, C.; Dimmel, D.; Schmidt, J.A. *Lignin and Lignans: Advances in Chemistry*; CRC Press, Tylor & Francis Group: Boca Raton, FL, USA, 2010.
53. Gordobil, O.; Egüés, I.; Llano-Ponte, R.; Labidi, J. Physicochemical properties of PLA lignin blends. *Polym. Degrad. Stab.* **2014**, *108*, 330–338. [CrossRef]
54. Al-Itry, R.; Lamnawar, K.; Maazouz, A. Reactive extrusion of PLA, PBAT with a multi-functional epoxide: Physico-chemical and rheological properties. *Eur. Polym. J.* **2014**, *58*, 90–102. [CrossRef]
55. Weng, Y.-X.; Jin, Y.-J.; Meng, Q.-Y.; Wang, L.; Zhang, M.; Wang, Y.-Z. Biodegradation behavior of poly(butylene adipate-co-terephthalate) (PBAT), poly(lactic acid) (PLA), and their blend under soil conditions. *Polym. Test.* **2013**, *32*, 918–926. [CrossRef]
56. Rahman, M.; Afrin, S.; Haque, P.; Islam, M.; Islam, M.S.; Gafur, M. Preparation and Characterization of Jute Cellulose Crystals-Reinforced Poly (L-lactic acid) Biocomposite for Biomedical Applications. *Int. J. Chem. Eng.* **2014**, *2014*, 842147. [CrossRef]
57. Lu, Q.; Liu, W.; Yang, L.; Zu, Y.; Zu, B.; Zhu, M.; Zhang, Y.; Zhang, X.; Zhang, R.; Sun, Z.; et al. Investigation of the effects of different organosolv pulping methods on antioxidant capacity and extraction efficiency of lignin. *Food Chem.* **2012**, *131*, 313–317. [CrossRef]

58. Mahmood, Z.; Yameen, M.; Jahangeer, M.; Riaz, M.; Ghaffar, A.; Javid, I. Lignin as Natural Antioxidant Capacity. In *Lignin-Trends and Applications*; IntechOpen: London, UK, 2018; pp. 181–205.
59. Gkartzou, E.; Koumoulos, E.P.; Charitidis, C.A. Production and 3D printing processing of bio-based thermoplastic filament. *Manuf. Rev.* **2017**, *4*. [CrossRef]

© 2019 by the authors. Licensee MDPI, Basel, Switzerland. This article is an open access article distributed under the terms and conditions of the Creative Commons Attribution (CC BY) license (http://creativecommons.org/licenses/by/4.0/).

MDPI
St. Alban-Anlage 66
4052 Basel
Switzerland
Tel. +41 61 683 77 34
Fax +41 61 302 89 18
www.mdpi.com

Materials Editorial Office
E-mail: materials@mdpi.com
www.mdpi.com/journal/materials

www.ingramcontent.com/pod-product-compliance
Lightning Source LLC
LaVergne TN
LVHW070545100526
838202LV00012B/391